Physics of Schottky Electron Sources

Physics of Schottky Electron Sources

Theory and Optimum Operation

Merijn Bronsgeest

Pan Stanford Publishing

Published by

Pan Stanford Publishing Pte. Ltd.
Penthouse Level, Suntec Tower 3
8 Temasek Boulevard
Singapore 038988

Email: editorial@panstanford.com
Web: www.panstanford.com

British Library Cataloguing-in-Publication Data
A catalogue record for this book is available from the British Library.

Physics of Schottky Electron Sources:
Theory and Optimum Operation

ISBN 978-981-4364-79-9 (Hardcover)
ISBN 978-981-4364-80-5 (eBook)

Printed in the USA

Contents

Preface

The Schottky electron source is the predominant emitter type in focused electron beam equipment, but although used extensively and satisfactorily, its properties are not fully understood. New developments and increasingly more stringent performance requirements ask for a better understanding: What is the best possible performance from a Schottky source for a given application? Which operating parameters are associated with that performance, and how stable is this "best" performance? These are the questions addressed in this book.

The content for this book is the result of my work as a PhD candidate in the research group Charged Particle Optics at Delft University of Technology in the Netherlands, and it is my pleasure to acknowledge the interaction with Greg Schwind, Lyn Swanson, Sean Kellogg, and Alan Bahm of FEI Beamtech in Hillsboro, Oregon, USA, throughout the years; the technical assistance of Jan de Looff, Frans Berwald, and Jacques Nonhebel of the research group Charged Particle Optics in Delft and of Ted Tessner of FEI Company; the opportunity given to me by Mike Lysaght and Greg Schwind of FEI Company to work with the sources research team of FEI Company in Hillsboro in May–July 2006; the courtesy of FEI Company for the use of some of its scanning electron microscopy (SEM) images in this book; the contributions to the work on practical brightness by the late Jim Barth of the research group Charged Particle Optics in Delft; and the support and guidance of my doctoral advisor Pieter Kruit, without whom this book would not exist.

Merijn Bronsgeest

Introduction

Electron beams are used for many applications. The most well-known example is probably the old-fashioned cathode ray tube (CRT) television, which changed our daily life. And an important step for science and technology was the invention of the electron microscope. Today's focused electron beam equipment uses electron beams mainly for imaging purposes, elemental characterization, and writing.

With respect to imaging or chemical analysis, electron irradiation of a sample produces a wealth of different signals for detection, each with specific information on the sample, for example:

- elastically back-scattered electrons;
- secondary electrons (ejected from the sample by inelastic scattering interactions with the beam electrons);
- Auger electrons (a beam electron removes an electron from the core of an atom, leaving a vacancy, an electron from a higher energy level falls into the vacancy, and the energy that is released is transferred to another electron, which is ejected from the atom with a characteristic kinetic energy);
- characteristic X-rays (a beam electron removes an electron from the core of an atom, leaving a vacancy, an electron from a higher energy level falls into the vacancy, and the energy that is released is transferred to a photon);
- light (a beam electron promotes an electron from the valence band to the conduction band (semiconductor sample), leaving a hole, and the electron recombines with a hole and emits a photon); and
- whether the sample is thin enough: transmitted electrons, which can be analyzed on changes in amplitude, phase, and/or energy with respect to the primary beam.

"Writing" with electrons is used, for example, to:

- locally functionalize a surface;
- create a charge pattern on the surface (for nanoxerography);
- locally expose a so-called "resist" layer on the surface (maskless lithography); and
- selectively deposit (part of) a precursor gas (electron beam–induced deposition [EBID]) (resistless nanolithography).

Crucial to any application is the electron source. The predominant emitter type in today's focused electron beam equipment, and the topic of this book, is the so-called Zr/O/W{100} Schottky electron source.

Schottky electron sources are used in many different systems of different companies (e.g., FEI Company*, Jeol Ltd., Hitachi Ltd., Carl Zeiss, KLA-Tencor Corporation, Tescan s.r.o., Phi [Physical Electronics Inc.], RIBER, Applied Materials). To name a few examples: the Titan (scanning) transmission electron microscope (FEI Company), the Nova NanoSEM scanning electron microscope (FEI Company), the JBX-6300FS Electron Beam Lithography System (Jeol), the SU-70 scanning electron microscope (Hitachi), the JXA-8500F Electron Probe Micro Analyzer (Jeol), the Mira II CS scanning electron microscope (Tecsan), the PHI 700 Scanning Auger Nanoprobe (Physical Electronics), and the microscopes with the Gemini column (Carl Zeiss).

Each of the applications puts its own specific demands on the properties of the irradiation. To give two examples, in electron energy loss spectroscopy (EELS) material-specific information is obtained by measuring the loss in energy of the primary beam electron upon transmission through the sample, and the primary beam electrons thus preferably all have exactly the same energy, while for high-resolution imaging it is desirable to be able to focus the electron beam down to a very small spot on the sample with still enough primary electrons in it to generate a detectable signal within a reasonable amount of time. The latter requires a high reduced brightness of the beam.

Over the past decades the Schottky source has been used as an electron emitter extensively and satisfactorily. Its properties, however, are not fully understood, and the ongoing quest for ever

---

*FEI Beam Technology Division, 5350 NE Dawson Creek Dr., Hillsboro, OR 97124, USA.

higher resolution and throughput, in combination with the advances made in the quality of electron optical systems (e.g., aberration correctors), put an increasing demand on the performance of the source with respect to, for example, brightness, energy spread, and geometrical stability. Examples are the recent interest in multibeam and multisource systems (Zhang & Kruit, 2007; Dokania *et al.*, 2008), the TEAM project (Dahmen *et al.*, 2009), and the Magellan^{TM} (Young, 2009).

Such developments require better knowledge of the relation between the operating parameters of the source and the properties of the beam at the sample, and the stability of the physical shape of the emitter. These are the topics addressed in this book.

The Schottky source is typically operated in a vacuum environment of ~10^{-9} mbar, and the electron emission is generated by heating the source and applying an electric field to it. Figure i.1 gives images of the Schottky emitter at different length scales.

Figure i.1 Schottky emitters at different length scales.

The emitter consists of a single-crystalline tungsten wire, about 1 mm long and 125 µm in diameter. One end of the wire is etched down to a tip with a diameter of about 1 µm. Halfway along the wire a reservoir of ZrO_x is attached, and at the other end the wire is spot-welded to a polycrystalline tungsten loop. This loop is fixed to two poles, which are embedded in a cylindrical ceramic base 1 cm in diameter. By running 2.0–2.5 A of current through the loop the emitter is heated through resistive heating and thermal conduction.

At a standard operating temperature of 1,800 K the emitter and part of the tungsten loop are glowing brightly.

The electric field is applied by biasing the emitter negatively with respect to an extractor: a metallic plate or cone, with a central aperture, usually at about 0.5 mm distance from the emitter tip. The potential difference between the emitter and the extractor is typically a few kilovolts. In the standard configuration the emitter is operated with an extractor and a suppressor.

The suppressor is a metal cap with a small hole. The emitter protrudes from the suppressor cap through the aperture for about 0.25 mm. It can be seen just peeking out from the cap in Fig. i.2. In operation the suppressor is biased negatively with respect to the emitter up to a few hundred volts to suppress unwanted electron emission from the part of the emitter that is inside the cap.

Figure i.2 Schottky emitter with and without a suppressor cap.

Two examples of a suppressor-emitter-extractor module are shown in Fig. i.3.

Characteristic for the Schottky emitter is the low work function of the planes with the {100} orientation. This is caused by the presence of zirconium and oxygen on the surface, which reduces the work function of the {100} planes considerably with respect to other crystallographic orientations. Zirconium and oxygen are present in the reservoir on the base of the emitter, and at a high operating temperature they can reach the {100} planes through diffusion.

The most important {100} plane is the facet on the tip end (Fig. i.1). This facet delivers the electrons that will end up in the final beam for applications. Most of the emission from a Schottky source is not used: the total emission from a Schottky emitter is typically up to a few hundred microamperes. The emission from the end facet is typically a few tens of microamperes. The beam current that is used in applications is cut out from the facet beam and typically contains of the order of a few picoamperes up to maybe microamperes.

Figure i.3 Examples of integrated source-suppressor-extractor modules (left: YPS AES-170 module; right: Denka TFE module).

The Schottky source has been studied by several investigators. In total about 50 articles have been published. Swanson *et al.* have done a great deal of the pioneering work from the late sixties to the mideighties in the area of Portland, Oregon. This area is still a location of active research: it is where FEI Company has concentrated its emitter research and development efforts. FEI Company is one of the few manufacturers of Schottky emitters (others are Denka and York Probe Sources Ltd.).

The first patent on the Schottky source was filed in 1964 as a result of the research in Oregon. Since then, about 15 US patents have been awarded, mainly to FEI Company (Oregon), Hitachi Ltd. (Tokyo), and Denka (Tokyo), although three early patents are from Wolfe, who worked for General Electric Company (New York) and Burroughs Corporation (Detroit).

During the work for this book there have been many discussions with FEI Company.

The structure of this book is as follows: in the first part of this book beam properties are addressed. Chapter 1 gives the electron emission theory that is used throughout this book to quantify the properties of emission from a surface. Chapter 2 addresses how the input parameters for emission theory relate to the practical case of a Schottky emitter in operation. In Chapter 3 we zoom in onto the emission from the end facet and how that arrives at the extractor, and in Chapter 4 we make the connection between facet emission and the final beam properties relevant to applications.

The second part, Chapter 5, is dedicated to the geometrical stability of the shape and consequences for the properties of the beam.

Finally, in the last part, Chapter 6, the focus is on achieving and maintaining maximum performance from the source for different applications.

The Schottky effect, the lowering of the potential barrier at the surface by applying a field to it, plays an important role in the emission properties of the Schottky electron source (Fig. i.4).

Figure i.4 Walter Hermann Schottky (1886–1976).

Chapter 1

Electron Emission from a Surface

This chapter discusses and compares the theories that are used in the remainder of this book to calculate the properties of the emission from the Schottky emitter surface. These properties are the current density and energy distribution of the emitted electrons as a function of the temperature, work function, and Fermi energy of the emitting material, and the field strength that is applied. The main reference work is the first chapter of the book by Modinos (1984).

1.1 The Potential Energy Barrier at a Surface

For Schottky electron sources electrons are extracted by increasing the temperature of the source and by applying a negative potential to the source with respect to an extractor. At 0 K and with zero bias electrons cannot escape the surface of the Schottky electron source because of the work function. The work function is the potential difference between the vacuum potential and the Fermi energy, or the work to remove an electron from the metal at 0 K. The potential difference exists because the free electrons in metals will spill out toward the vacuum from the surface and the spill-out electrons and the positive charges remaining inside the metal produce an electric double layer (Lang & Kohn, 1971). The potential energy barrier at the surface is not a step function. An electron that has escaped the metal will experience just outside the metal a force that pulls the

Physics of Schottky Electron Sources: Theory and Optimum Operation
Merijn Bronsgeest

ISBN 978-981-4364-79-9 (Hardcover), 978-981-4364-80-5 (eBook)
www.panstanford.com

electron back to the metal. This force originates from the positive screening charge inside the metal. This gives a potential energy that depends on the distance to the metal–vacuum interface. The classical screening energy (image charge energy) for a perfectly conducting semi-infinite metal is equal to the work done in bringing a point charge from infinity (in the vacuum) to $-z$ (the distance between the surface and the screening charge, equal to distance between the surface and the point charge). The potential energy varies in space according to (in SI units)

$$U(z) = E_F + \phi + \int_{-z}^{\infty} \frac{-e^2}{4\pi\varepsilon_0 (2z)^2} dz = E_F + \phi - \frac{e^2}{16\pi\varepsilon_0 z}, \tag{1.1}$$

where E_F is the Fermi energy, ϕ the work function, and ε_0 the permittivity of vacuum. Figure 1.1 gives a total potential energy diagram for the metal–vacuum interface (in eV) that shows the work function and the classical image potential.

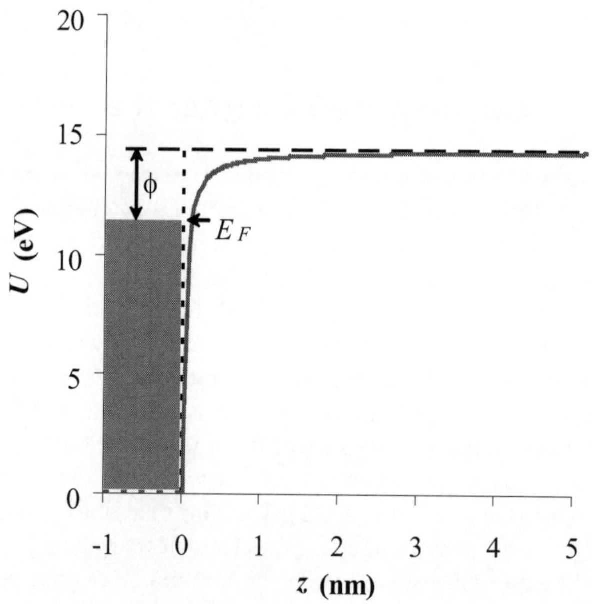

Figure 1.1 Potential energy diagram for the metal–vacuum interface showing the classical image potential (solid curve).

The image potential ignores the atomic structure of the metal surface and is generally assumed to be a valid approximation of the potential for distances greater than the lattice parameter. Traditionally the image potential is truncated (horizontal dashed line in Fig. 1.1 at $U = 0$ eV). Nonclassically, there are small oscillations in the potential in the metal near the surface, even for a free-electron metal. The oscillations are associated with charge density oscillations ("Friedel" oscillations) in the direction perpendicular to the surface of a crystal that arise due to the breaking of the translational symmetry at the surface (Lang & Kohn, 1970). Because of these oscillations there is a weak reflecting region at the metal–vacuum interface, which will be addressed in more detail later (Section 1.4.1).

1.2 Emission by Heating

One of the mechanisms to get electrons to overcome the potential barrier and escape from the surface is to provide them with enough energy by heating. We will discuss this mechanism using the free-electron theory of metals. (It is noted that tungsten, the main Schottky emitter material, actually is a transition metal with localized d electrons.)

In the free-electron theory of metals electron states are described by plane waves with wave vectors $\mathbf{k}$. The energy of an electron in a state with wave vector $\mathbf{k}$ is

$$E = \frac{\hbar^2 \mathbf{k}^2}{2m}. \tag{1.2}$$

The number of electron states per unit volume that have an energy between E and $E + dE$ is

$$\begin{aligned} \rho(E)dE &= \frac{2}{(2\pi)^3} \iiint_E^{E+dE} d^3\mathbf{k} = \frac{2}{(2\pi)^3} 4\pi \mathbf{k}^2 d\mathbf{k} \\ &= \frac{1}{2\pi^2} \left(\frac{2m}{\hbar^2} \right)^{3/2} \sqrt{E}\, dE. \end{aligned} \tag{1.3}$$

The probability of an electron state with energy E being occupied is (Fermi–Dirac statistics)

$$f(E) = \frac{1}{1 + e^{(E - E_\text{F})/k_\text{B}T}}. \tag{1.4}$$

At 0 K the Fermi energy is the highest of the energies of the electron states in the metal that can be occupied. At finite temperatures states with higher energies can become occupied, but at room temperature the occupation probability for an energy level as high as the potential barrier at a surface is still very small and the electron escape probability is generally negligible. At 1,800 K (the Schottky emitter operating temperature) the occupancy of an energy level of, for example, 2.9 eV above the Fermi level (the Schottky emitter work function) has reached $\sim 10^{-8}$.

The next step in calculating the emission is to calculate the number of electrons arriving at the surface, per unit time and area, from the inside of the metal. We take the surface parallel to the *xy*-plane and the positive *z*-direction going out from the surface and thus select all electrons with $v_z > 0$. The normal (perpendicular to the surface) energy of the electrons is given by

$$W = \frac{\hbar^2 k_z^2}{2m} = E - \frac{\hbar^2 k_{||}^2}{2m}. \tag{1.5}$$

The number of electrons per unit time impinging on the metal–vacuum interface with total energy between E and $E + dE$ and normal energy between W and $W + dW$ is

$$N(E,W)dEdW = \frac{2f(E)}{(2\pi)^3} \iiint_{(E,W)} v_z d^3\mathbf{k}, \tag{1.6}$$

with the normal velocity v_z by

$$v_z \equiv \frac{1}{\hbar}\frac{\partial E_\mathbf{k}}{\partial k_z} = \frac{\hbar \mathbf{k}_z}{m}. \tag{1.7}$$

And this can be rewritten to give

$$N(E,W)dEdW = \frac{m}{2\pi^2\hbar^3} f(E)dEdW. \tag{1.8}$$

The total number of electrons impinging on the surface per unit time per unit area with normal energy between W and $W + dW$ is found by integrating over all total energies:

$$N(W)dW = \frac{m}{2\pi^2\hbar^3} dW \int_W^\infty f(E)dE = \frac{m k_B T}{2\pi^2\hbar^3} \ln\left(1 + e^{-\frac{W-E_F}{k_B T}}\right) dW. \tag{1.9}$$

Not all of those electrons can escape from the surface: each normal energy has its own "transmission probability" D. The emitted current density, or the number of electrons per unit time per unit area, multiplied with their charge, is found from

$$j = e\int_0^\infty N(W)D(W)dW \,. \tag{1.10}$$

These emitted electrons will display a distribution of energies. The normal energy and tangential energy of an electron are conserved upon emission. The energy distribution of the emitted electrons is relevant to practical applications of electron sources, for example, because the effect of the electrostatic and/or magnetic lenses in the electron-optical system will be different for electrons of different energies. The general equations for the normal, tangential, and total energy distributions are given below for completeness. These will be applied to the specific case of emission from a heated surface in a field in Section 1.4.3.

The normal energy distribution of the emitted electrons can be found by multiplying Eq. (1.9) with the transmission probability

$$\begin{aligned} j(W)dW &= N(W)D(W)dW \\ &= \frac{me}{2\pi^2\hbar^3}\mathrm{k_B}T\ln\left(1+e^{-\frac{W-E_\mathrm{F}}{\mathrm{k_B}T}}\right)D(W)dW. \end{aligned} \tag{1.11}$$

The tangential energy distribution can be found by integrating the product of the Fermi–Dirac distribution and the transmission probability over all normal energies

$$j(E_{\tan})dE_{\tan} = \frac{me}{2\pi^2\hbar^3}dE_{\tan}\int_0^\infty \frac{D(W)}{1+e^{(E_{\tan}+W-E_\mathrm{F})/\mathrm{k_B}T}}dW \,. \tag{1.12}$$

The total energy distribution can be found from

$$j(E)dE = \frac{me}{2\pi^2\hbar^3}\frac{1}{1+e^{(E-E_\mathrm{F})/\mathrm{k_B}T}}\int_0^E D(W)dW \,. \tag{1.13}$$

To estimate the amount of thermal emission we assume the transmission function D is a classical step function: all electrons with energies lower than the Fermi energy plus the work function cannot escape, and all electrons with higher energies can. Nonclassically, the transmission is not a perfect step function due to reflection of the electron waves. This will be addressed in Section 1.4.1. Using the step function for the transmission the current density emitted by the heated surface is

$$j_{\text{thermal}}(T)=e\int_0^\infty N(W,T)D(W)dW$$
$$=e\int_{E_F+\phi}^\infty \frac{mk_BT}{2\pi^2\hbar^3}\ln\left(1+e^{-\frac{W-E_F}{k_BT}}\right)dW\approx\frac{em(k_BT)^2}{2\pi^2\hbar^3}e^{\frac{-\phi}{k_BT}}. \quad (1.14)$$

For the last step in the equation above the approximation $\ln(1 + e^{-x}) \approx e^{-x}$ was used because most of the emission comes from near the top of the potential barrier for which $(W - E_F)$ is much larger than k_BT.

As a quantitative example, the current density from a surface heated to 1,800 K is 1 A/m^2 if the work function is 4.5 eV (bare tungsten) and 30 kA/m^2 if the work function would be 2.9 eV (Schottky emitter). The heating thus induces electron emission, and it is noted that a favorable side effect is that it also reduces the contamination level of the surface, making the vacuum requirements for the electron source to work less stringent.

1.3 The Effect of an Electric Field on the Potential Energy Barrier at a Surface

In standard operation a Schottky electron source is heated, but it is also biased with respect to an extractor. By applying an electric field the thermal current density given above can be increased considerably. The typical gun geometry for the Schottky emitter, with the suppressor and the extractor, has been described in the introduction. The bias between the source and the extractor changes the potential barrier for electrons at the source surface and thus affects the emission properties. We first calculate the variation of the electrical potential in the gun area and then look at the consequences for the potential energy diagram of Fig. 1.1.

The potential and/or field distribution in the electron gun can be accurately calculated with a charge density or boundary element method (http://www.electronoptics.com/). In this method the applied potentials are assigned to the contours of the electrodes, and the software then finds the associated charge distribution across the surface of the electrodes. Next, the electrodes are effectively replaced by the calculated surface charges, which form the input for evaluations of the potential or field at any desired position.

Figure 1.2 shows a potential contour that corresponds to that for a typical gun geometry for a Schottky emitter, followed by a field-free zone. The gap between the emitter and the suppresser, and between the suppressor and the extractor, is bridged by an electrode on which the potential is varied linearly. For the gap between the emitter and the suppressor the linear variation is not a very good approximation, but it is has been verified that it will not affect the field strength at the emitter tip. Alternatively, one could extend the design and use the true geometry of the suppressor and leave open the gap between the emitter and the suppressor.

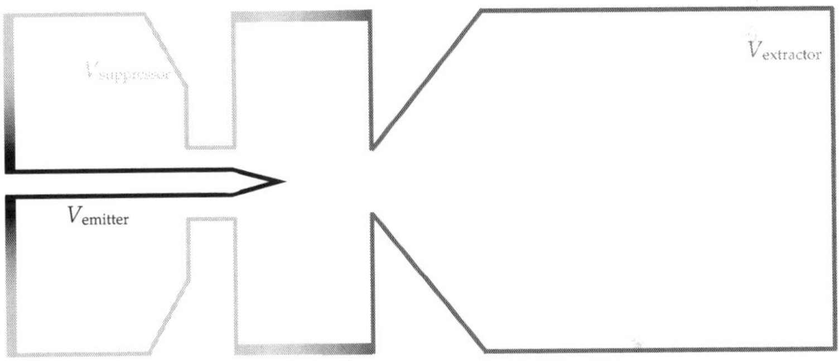

Figure 1.2 Example of the typical Schottky gun geometry with the imposed potential contour that is used for the charge density method calculation. The gun geometry is the standard gun geometry with a suppressor cap and the extractor at 0.76 mm from the suppressor.

Figure 1.3 gives a typical distribution of equipotential lines in the section between the suppressor and the extractor. While in principle the potential can be calculated accurately at any point, the equipotential lines follow from the evaluation of the potential for a chosen number of grid points.

Figure 1.3 shows the field strength between the suppressor and the extractor plate is approximately constant (3.3/0.75 kV/mm), while the field is enhanced at the sharp tip end. We will have a closer look at the general consequences of the distribution in Fig. 1.3 in Section 2.2 but focus now on the variation in field potential and field strength along the optical axis. This is shown in Fig. 1.4 for a suppressor voltage of −0.3 kV and an extraction voltage of 5 kV with respect to the tip. Also given is the potential and field strength variation in the first 100 nm measured from the tip.

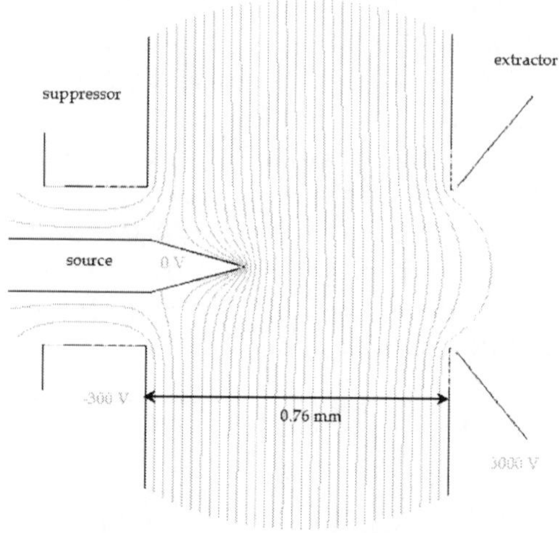

Figure 1.3 Equipotential lines between the suppressor and the extractor for V_{sup} = −0.3 kV and V_{ext} = 3.0 kV with respect to the emitter (equipotential lines in steps of 100 V).

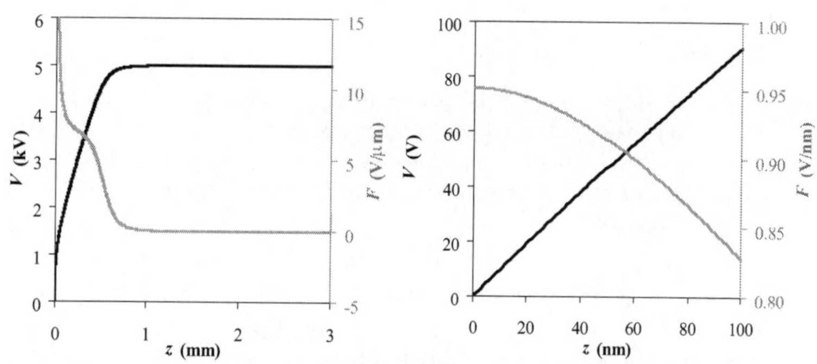

Figure 1.4 Electric potential V and electric field strength F along the optical axis of the geometry in Fig. 1.2, starting at the emitter tip end, for V_{sup} = −0.3 kV and V_{ext} = 5 kV with respect to the emitter. The dashed line indicates the position of the extractor.

By applying a bias between the emitter and the extractor a potential difference is created between the Fermi level of the extractor and the Fermi level of the emitter. In the example of Fig. 1.4 the potential energy of the electrons inside the metal is raised with 5 keV with respect to the extractor. To construct a new potential energy diagram for the electrons, the field potential of Fig. 1.4 has to be added to the image potential of Fig. 1.1. For emission calculations the field potential in Fig. 1.4 is generally simplified. Figure 1.4 shows that the electric field strength in the first ~20 nm is approximately constant (and the potential is approximately linear), as in a parallel-plate situation. This is the general assumption that is made to get an analytical description of the potential barrier near the surface with (in SI units)

$$\begin{aligned} U(z) &= E_{\mathrm{F}} + \phi + \int_{-z}^{\infty} \frac{-e^2}{4\pi\varepsilon_0 (2z)^2} dz + e\int_{0}^{z} -F dz \\ &= E_{\mathrm{F}} + \phi - \frac{e^2}{16\pi\varepsilon_0 z} - eFz. \end{aligned} \tag{1.15}$$

Figure 1.5 shows the resulting potential energy diagram (in eV) close to the surface of a Schottky emitter in the configuration of Fig. 1.2. It is noted that the constant field approximation will be increasingly less good when the field is lower, but it has been verified that the effect on the calculated current density emitted by Schottky emitters is negligibly small.

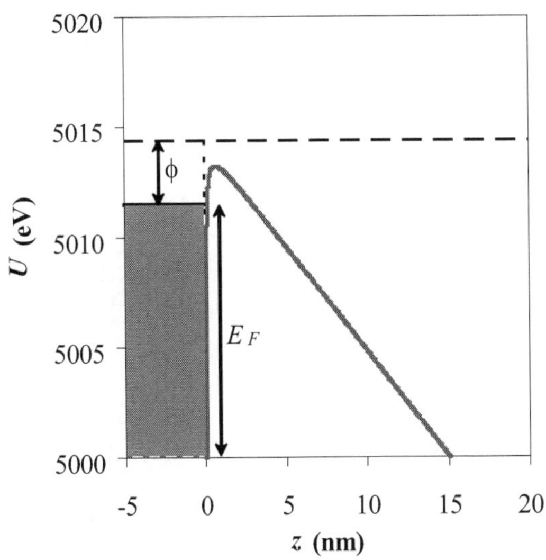

Figure 1.5 Analytical description of the potential energy barrier near the surface of a perfectly conducting semi-infinite metal for a field strength of 0.95 V/nm.

As can be seen in Fig. 1.5 the field lowers the height of the potential barrier at the surface with respect to the zero-bias case.

This work function reduction is the so-called Schottky effect. The Schottky effect decreases the potential barrier maximum to

$$U_{\max} = E_{\mathrm{F}} + \phi - \sqrt{\frac{e^3 F}{4\pi\varepsilon_0}} \,. \tag{1.16}$$

1.4 Emission by Heating and Applying an Electric Field

1.4.1 Escape Probability

To calculate the amount of emission by heating in the presence of an electric field we need the transmission probability $D(W)$ in Eq. (1.10) in the presence of the field. The field-dependent Schottky effect lowers the barrier, which allows for more electrons to escape, but in addition to that, if the field is strong enough, the top of the potential barrier becomes thin enough for electrons to tunnel through. The basic requirement for the nonclassical transmission calculation is that at each point in space the wave function of the electron has to satisfy the Schrödinger equation. Because the potential energy changes only normal to the surface, the wave function can be written as

$$\psi = e^{i\mathbf{k}_{||}\mathbf{r}_{||}} \varphi(z) \,, \tag{1.17}$$

where $\mathbf{r}_{||}$ is the position vector and $\mathbf{k}_{||}$ the wave vector in the plane parallel to the surface and $\varphi(z)$ is the component that has to satisfy the Schrödinger equation:

$$-\frac{\hbar^2}{2m}\frac{d^2\varphi}{dz^2} + U(z)\varphi = W\varphi \,. \tag{1.18}$$

For a *constant* potential energy U we know solutions to the Schrödinger equation:

$$\varphi(z) = C_1 e^{i\mathrm{k}_z z} + C_2 e^{-i\mathrm{k}_z z} = C_1 \cos \mathrm{k}_z z + C_2 \sin \mathrm{k}_z z \ \text{ for } W > U, \tag{1.19}$$

$$\varphi(z) = C_3 e^{\mathrm{k}_z z} + C_4 e^{-\mathrm{k}_z z} \ \text{ for } W < U, \tag{1.20}$$

with constants C_1, C_2, C_3, and C_4.

In a potential *gradient* like that of Fig. 1.5 the wave vector must vary from point to point to keep the total energy $W + U(+E_{||})$ constant. To find the transmission probability D, the next approximation is that we assume that the probability for an electron to get from the emitter surface to the extractor is equivalent to the probability for an electron to get through the barrier given in Fig. 1.6 below. This means that we assume there is no reflection going from z_2 to the extractor.

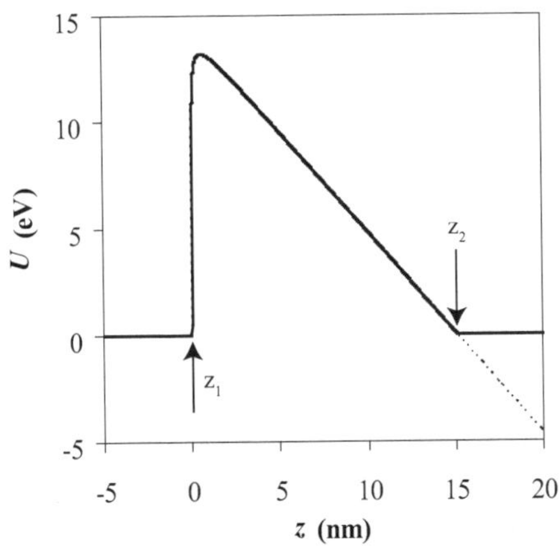

Figure 1.6 Potential energy barrier for the calculation of the nonclassical transmission probability for an electron to get from the emitter surface to the extractor. The reference energy is the lowest energy level in the metal. The dashed line indicates the "true" barrier that continues to decline toward the extractor.

The potential is constant to the left and right of the barrier in Fig. 1.6, which means we have solutions, Eqs. (1.19) and (1.20), to the Schrödinger equation for both sides. Also, the potential has the same value on both sides, which means the wave vector is the same on both sides. The unknowns are the constants. For the left side we have an incident wave and a reflected wave, and on the right side we have the transmitted wave:

$$\text{For } 0 < z < z_1\text{: } \varphi(z) = C_1 e^{ik_z z} + C_2 e^{-ik_z z}, \tag{1.21}$$

$$\text{and for } z > z_2\text{: } \varphi(z) = C_3 e^{ik_z z}. \tag{1.22}$$

The transmission probability is now given by

$$D(W) = \left|\frac{C_3}{C_1}\right|^2. \tag{1.23}$$

Because for the transmission probability we need the ratio of the two constants we can set one to an arbitrary value. By setting C_3 to 1 we have the complete solution for the right side of the barrier, and we have the requirement that the wave function has to be continuous and smooth everywhere. By numerically integrating the Schrödinger equation, moving through the barrier from right to left in a step-by-step procedure, the value of the wave function and its derivative can be found at the other side of the barrier at $z = z_1$, where the wave function has to match with a plane-wave equation again. This requirement will yield the value of C_1.

Figure 1.7 shows an example for the calculated transmission for two different field strengths. Also included is the classical stepped transmission function.

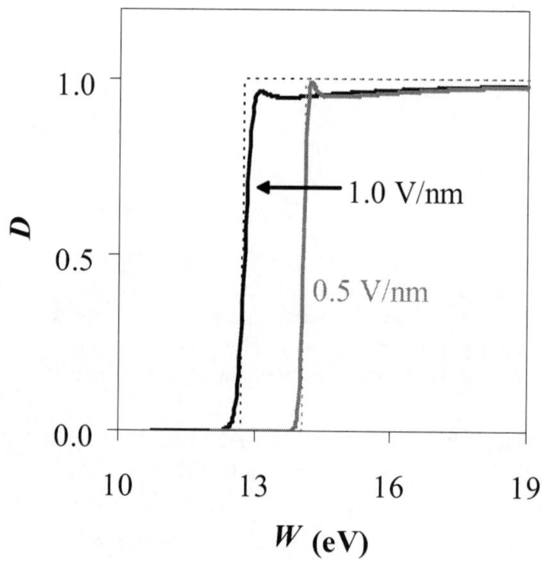

Figure 1.7 Transmission as a function of normal energy, calculated for the potential barrier of Eq. (1.12) for two different field strengths and with $\phi = 2.93$ eV and $E_F = 11$ eV. Dashed lines indicate the classical transmission functions.

Figure 1.7 shows that tunneling is possible for electrons with energies below the barrier peak and also that the transmission probability above the barrier is not 100%. The tunneling contribution is an important factor for a Schottky emitter in normal operating conditions.

The differences in the shape of the two nonclassical transmission functions in Fig. 1.7 can be ascribed to a different reflection response of the electron waves for the different potential barriers.

If the potential barrier would have been joined smoothly to the regions of constant potential to the left and right, reflection of electrons with energies below the barrier would occur only between the z-positions for which the electron energy equals the potential function and for electrons with energies above the barrier, from the immediate vicinity of the barrier peak. But because of the abrupt changes in the potential of Fig. 1.6, there is also reflection at $z = z_1$ and $z = z_2$. The abrupt potential change near the surface approximately simulates "real" reflection and interference that exists at a metal–vacuum interface that originate from the local charge density oscillations. The reflection for electrons with energies near or above the barrier peak takes place roughly between z_1 and $z_{\max}$ and $z_{\max}$ and z_2. When the distance between z_1 and $z_{\max}$ or $z_{\max}$ and z_2 is approximately equal to an integral multiple of the average electron wavelength, constructive interference can take place, and this causes variations in the transmission as a function of electron energy. The reflection effect depends on the position and height of the barrier maximum with respect to z_1 and z_2 and will thus be different for a different field strength at the surface and for a different Fermi energy. Figure 1.8 shows part of the transmission function calculated for different Fermi levels at a field strength of 1 V/nm.

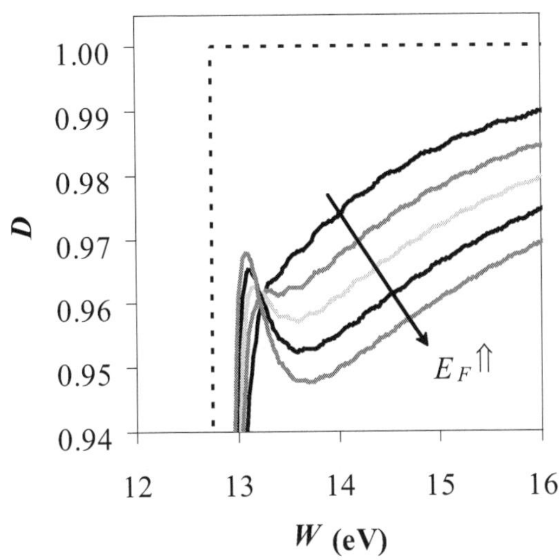

Figure 1.8 Effect of the Fermi energy on the calculated transmission function for F = 1.0 V/nm and ϕ = 2.93 eV. The Fermi energy was varied from 4 eV to 12 eV.

For convenience there are analytical approximations available to get a transmission function that includes the tunneling. See for a recent discussion with a focus on cold field electron emission (Forbes, 2008). For emission from Schottky emitters the approximations generally include a parabolic barrier approximation for the potential function and the Jefferson–Wentzel–Kramers–Brillouin (JWKB) approximation (e.g., Griffiths, 2004) for finding the solution to the Schrödinger equation. The inspiration behind the JWKB approximation is that when the potential varies rather slowly in comparison to the electron wavelength (for $W > U$), the wave function remains sinusoidal, except that the wavelength and amplitude change slowly with z. (Similarly, for $W < U$, when the potential varies slowly in comparison to the decay length of the wave function, the wave function remains practically exponential.) These assumptions allow one to find an analytical solution of the Schrödinger equation. It goes wrong near points for which $W = U$, but near such "turning points" the Schrödinger equation is solved by approximating the potential function with a straight line, and that solution is then patched to the wave functions before and after the turning point.

For the general case of a parabolic potential barrier, Kemble (1935) has shown that the JWKB approach yields for the transmission function

$$D(W) = \frac{1}{1+e^{G}}, \tag{1.24}$$

with

$$G(W) = -\frac{2i}{\hbar}\int_{z_1}^{z_2} \sqrt{2m(W-U(z))}\,dz\,. \tag{1.25}$$

z_1 and z_2 are the solutions for which $W - U(z) = 0$. Note that for $W < U_{max}$, z_1 and z_2 are real, but for $W > U_{max}$, z_1 and z_2 are complex. Christov (1966) has used this result to get a transmission function for electron emission from a surface for which the potential barrier was given by Eq. (1.15). In his approach the potential barrier of Eq. (1.15) is expanded in a Taylor series around the barrier peak position:

$$U_{pb}(z) = U_{max} - \frac{e^2}{16\pi\varepsilon_0 z_{max}^3}(z - z_{max})^2\,. \tag{1.26}$$

An example of this approximation is shown (in eV) in Fig. 1.9.

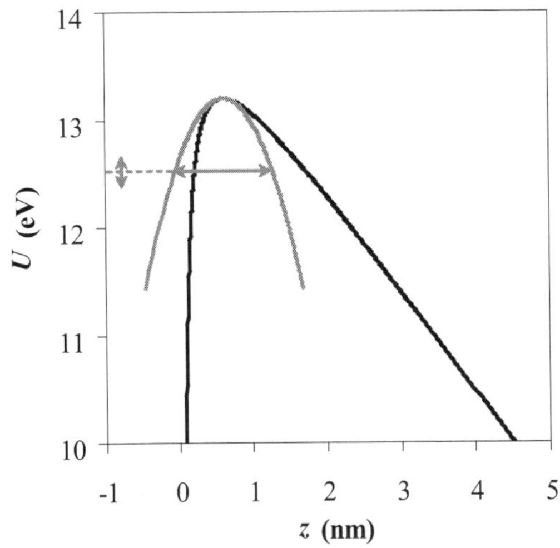

Figure 1.9 Parabolic barrier approximation of the potential energy barrier at the metal–vacuum interface for F = 0.96 V/nm. The horizontal arrow indicates the z-range in which the barrier approximation is applied to calculate the transmission coefficient for energy U analytically.

The integration range z_1 to z_2 (the horizontal arrow in Fig. 1.9) in Eq. (1.25) depends on the electron energy (as is illustrated by the vertical arrow in Fig. 1.9). For the parabolic barrier in Eq. (1.26), Eq. (1.25) can be integrated to give

$$G(W)=\frac{2\pi^2\sqrt{m}}{h(4\pi\varepsilon_0 e)^{1/4}}\frac{U_{\max}-W}{F^{3/4}}. \tag{1.27}$$

For $W = U_{\max}$ Eq. (1.27) becomes zero, and the transmission factor, from Eq. (1.24), is 0.5. For $W > U_{\max}$ Eq. (1.27) becomes negative, and for large W the transmission factor, Eq. (1.24), approaches 1.

Figure 1.10 compares the transmission functions of the analytical approximations with and without tunneling, and the numerical solution.

Figure 1.10 shows that the analytical transmission function that includes the tunneling exceeds the numerically calculated transmission function for all energies. The analytical transmission function with tunneling ignores reflection at the metal–vacuum interface.

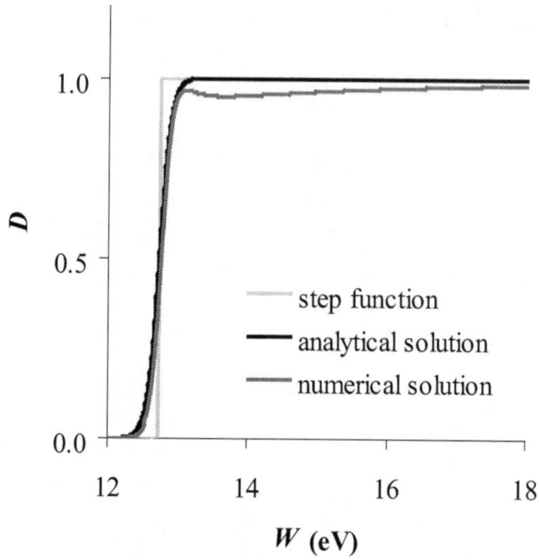

Figure 1.10 Comparison of the transmission functions for F = 1 V/nm, ϕ = 2.93 eV, and E_F = 11 eV.

1.4.2 Current Density

The previous section has yielded different methods to find the transmission function $D(W)$ as a function of field strength. Now we can determine the electron emission from a heated surface in the presence of a field. Using the classical transmission function that includes the Schottky effect but ignores tunneling, the current density is given by (Schottky emission)

$$j_S(F,T) = \frac{em}{2\pi^2\hbar^3}(k_B T)^2 e^{-\frac{\phi - \sqrt{\frac{e^3 F}{4\pi\varepsilon_0}}}{k_B T}}, \qquad (1.28)$$

for which, as before, the Fermi–Dirac distribution has been approximated with the Maxwell–Boltzmann distribution by omitting the unit term in the denominator of Eq. (1.4).

When there is a contribution to the emission from tunneling through the very top of the potential barrier, Murphy and Good (1956) have derived a modified version of the Schottky emission equation. The calculation is based on the analytical transmission function of Eq. (1.24) and is a reasonable approximation only for specific combinations of field, temperature, and work function.

Within the limits of its applicability, the current density can be given by (extended Schottky emission)

$$j_{\mathrm{ES}}(F,T)=j_{\mathrm{S}}\frac{\pi q}{\sin \pi q} \text{ with } q=\frac{\hbar}{\pi \mathrm{k_B} T\sqrt{m}}(4\pi\varepsilon_0 eF^3)^{1/4}. \tag{1.29}$$

Again the unit term in the denominator of the Fermi–Dirac distribution has been neglected. Note that the equation will diverge when q approaches unity. Typically $q = 0.7$ is taken as an upper limit (Fransen, 1999; Swanson & Schwind, 2008).

Figure 1.11 compares the result for the analytical equations for Schottky emission and extended Schottky emission, with the numerically calculated current density.

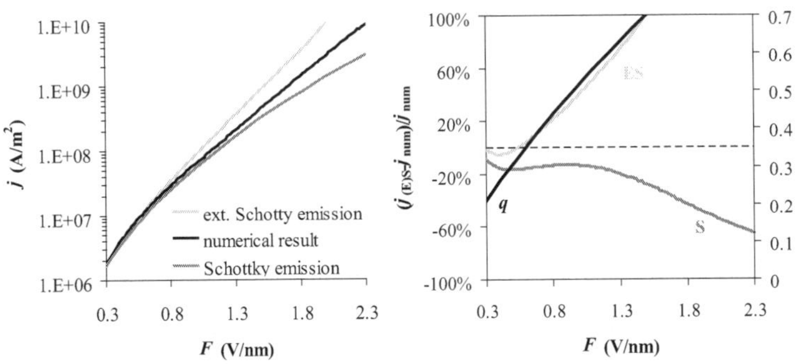

Figure 1.11 Current density as a function of electric field strength (left) for different calculation methods. T = 1,800 K, ϕ = 2.93 eV, and for numerical calculations E_F = 11.47 eV. (Right) Difference between extended Schottky emission (ES) and the numerical calculation and the Schottky emission equation (S) and the numerical result.

Figure 1.11 suggests the upper limit for q in Eq. (1.29) should be less than 0.7, perhaps <0.4, for the extended Schottky emission equation to be a valid approximation. Recently Bahm *et al.* (2008) have suggested to change the upper limit to 0.25.

Figure 1.12 zooms in onto the difference between the numerical result and the result from the Schottky emission equation for field strengths near 1 V/nm and below.

The periodic fluctuations in the deviation from the "Schottky line" are the result of the electron wave reflection and interference for the barrier shape of Fig. 1.6, which to some extent mimics the real reflection and interference at a metal-vacuum interface, associated with the charge density oscillations in the metal near the surface. The periodic fluctuations around the Schottky line have been demonstrated experimentally, for example, for an emitting W(111) surface (Stafford & Weber, 1963). One could thus experimentally verify the assumptions in the theory. This will be further investigated in Section 3.4.

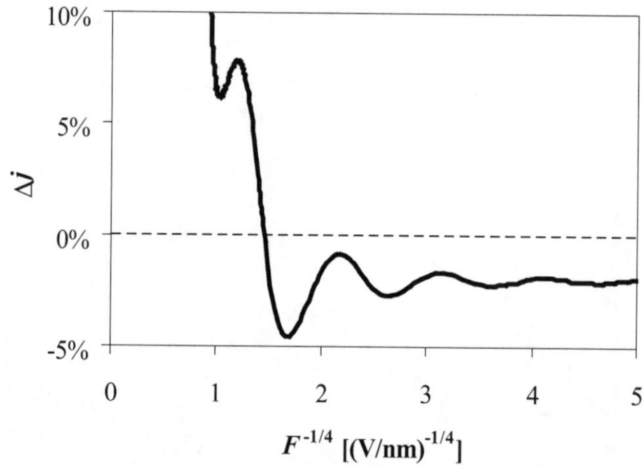

Figure 1.12 Deviation of the numerically calculated current density from the analytically calculated current density (neglecting tunneling). E_F = 11.47 eV, ϕ = 2.93 eV, and T = 1,800 K.

Finally, it is noted that the (dis)agreement between the analytical (extended) Schottky emission equation and the numerical solution depends on the chosen work function and temperature and also on the Fermi energy chosen for the numerical calculation, as can be anticipated from Fig. 1.8. In conclusion, to calculate the number of electrons emitted by a surface requires quite a few assumptions (free electron metal, barrier shape, etc.), and the work function, temperature, and field strength at the surface (which are not always easy to determine) and the outcome also depend on the calculation method. Which theoretical method is closest to experimental results remains to be seen. It would be useful for comparisons if in general one would specify the calculation method in detail.

In the remainder of this book both the analytical and the numerical approach will be used. The analytical equations have been used mainly in the earlier work described in this book. If not stated, we will use a numerically calculated current density using the potential barrier shape of Eq. (1.15) and truncate it, as shown in Fig. 1.6. For the Fermi energy we will take 11.47 eV (Petroff & Viswanathan, 1971).

1.4.3 Energy Distributions

The energy distribution of the emitted electrons is important to many applications of Schottky sources, such as for high-resolution imaging at low beam energies, but also for imaging at high beam energies in systems with spherical aberration correctors and, for example, electron energy loss spectroscopy. If the electrons display a distribution of energies it broadens the spot size of a focused beam and limits the temporal beam coherence of collimated beams due to chromatic lens aberrations. Figure 1.13 illustrates the origin of the typical emitted energy distribution: for energies below the barrier peak the occupation of electron energy levels inside the metal is high but the transmission probability low, while for energies above the barrier the escape probability is high but the occupation of the energy levels inside the metal low. Most emitted electrons have an energy close to the barrier maximum.

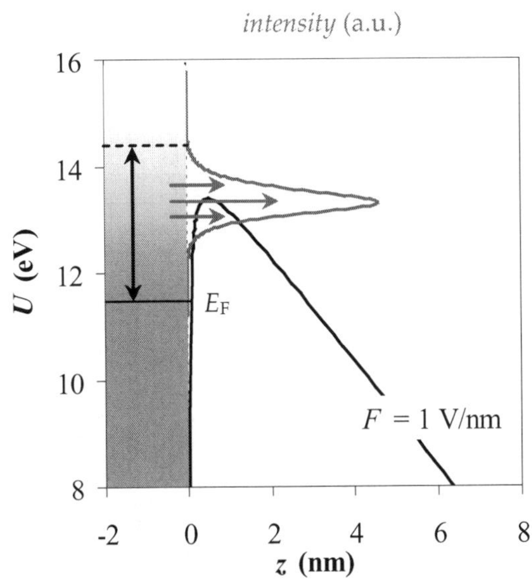

Figure 1.13 Principle of electron emission from a hot metal surface subjected to a field of 1 V/nm, illustrating the origin of the finite energy distribution of the electrons.

Figure 1.14 gives an example of the total energy distribution emitted by a surface at 1,800 K, subjected to a field strength of 0.5 V/nm, 0.8 V/nm, and 1.0 V/nm. The current density corresponds to the area under the curve and amounts to 6.5 × 10^7 A/m^2, 2.9 × 10^7 A/m^2, and 0.7 × 10^7 A/m^2 in order of decreasing field strength.

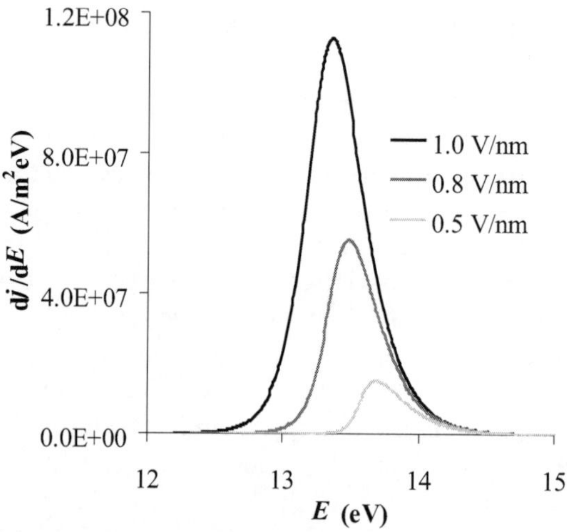

Figure 1.14 Total energy distribution of electrons emitted by a surface at 1,800 K, with a work function of 2.93 eV, calculated from Eq. (1.13).

Figure 1.14 shows that with increasing field strength the peak of the energy distribution shifts to lower energies and that the shape changes due to the increasing contribution from tunneling. Figures 1.15 and 1.16 give the associated normal and tangential energy distributions.

The average normal energy of the electrons upon emission is generally negligible with respect to the typical beam energies at the extractor in an operating Schottky source module, and it will be ignored in the following, unless stated otherwise.

The tangential energy distributions in Fig. 1.16 can be very well approximated with Maxwell–Boltzmann distributions with a mean of $k_B T$. For the standard operating temperature of 1,800 K the mean tangential energy is 0.155 eV. It is noted that the exponential function is a reasonable approximation also for cold field emitters, with a mean tangential energy of $d = e\hbar F/\sqrt{(8m\phi)}$ (Worster, 1969; El-Kareh *et al.*, 1977). These approximations will be used in Chapter 4.

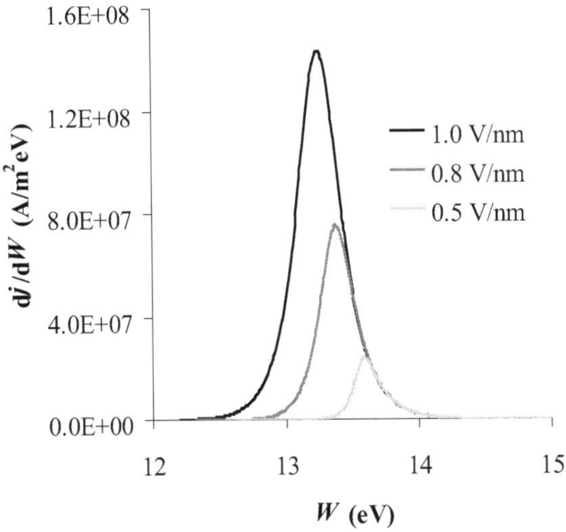

Figure 1.15 Normal energy distribution of electrons emitted by a surface at 1,800 K, with a work function of 2.93 eV, calculated from Eq. (1.12).

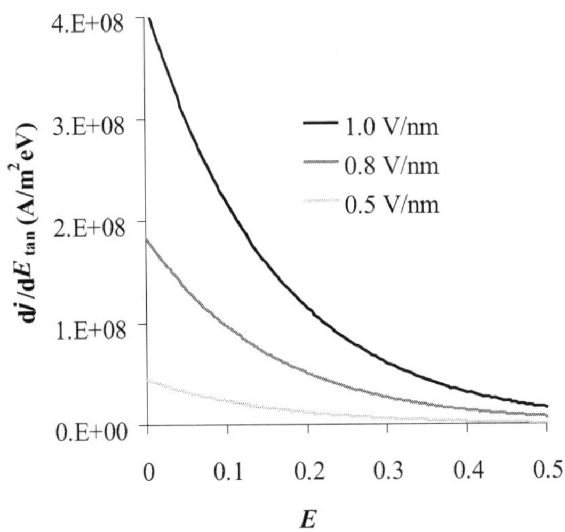

Figure 1.16 Tangential energy distribution of electrons emitted by a surface at 1,800 K, with a work function of 2.93 eV, calculated from Eq. (1.13).

To conclude, the emission theory in this chapter, for infinite metal surfaces, can give the current density and energy distribution of the emitted electrons as a function of the temperature and field. How this applies to the case of a real Schottky emitter geometry is addressed in the next chapters.

Chapter 2

Emission from a Schottky Emitter

For a practical application, basically only the electrons emitted by the (central part of the) end facet of a Schottky source are of interest, as these end up in the final beam that illuminates the specimen. But before zooming in we first take a broader look and consider the total emission from a Schottky source. The total emission from a Schottky source is relevant, for example, the gun design, because it could cause unwanted charging or heating of parts, which could damage the system. Although the content of this chapter has not been an area of thorough investigation, it is presented up front to sketch the general picture of a working Schottky emitter.

How much current a Schottky source emits can be calculated with the theory of the previous chapter if the field, temperature, and work function across the emitter surface are known. The work function is basically a material property, but the field and temperature are applied by biasing the emitter and by resistive heating. Below we will address the relation between the operating parameters for a Schottky emitter (bias, heating current) and the input parameters for emission theory (field and temperature at the emitting surface). The temperature is relatively constant across the emitter cone (McGinn, 2005), but as will be shown, the field and work function strongly vary. As a consequence, the current density distribution is a strong function of the position on the emitter, and the total emission is to be found from a surface integral.

Physics of Schottky Electron Sources: Theory and Optimum Operation
Merijn Bronsgeest

ISBN 978-981-4364-79-9 (Hardcover), 978-981-4364-80-5 (eBook)
www.panstanford.com

2.1 Work Function Variations across the Emitter Surface

The work function of a particular surface will depend on the arrangement of the ion cores and electron sea around it and is thus a function of the crystallographic orientation. A Schottky source consists primarily of single-crystalline tungsten, and there is a wide spread in theoretical and experimental work function data for tungsten surfaces of different crystallographic orientations. Overviews are given by, for example, Kawano (2008) and in the Landoldt–Börnstein database. For bare tungsten the work function for polycrystalline material is 4.5 eV, and the work function difference between two orientations has been reported to be as much as 1 eV. The end facet of a Schottky electron source, which is relevant to the electron beam used in applications, is a W{100} plane. The reported work function for W{100} is 4.6 eV, but the Schottky source has the ZrO_x reservoir, from which ZrO_x complexes can diffuse and cover the emitter surface and change the work function. For the W{100} surface the work function reduction is spectacular: the end facet of an operating Schottky source has a work function closer to 3 eV.

In the literature there is some controversy on how exactly the ZrO_x is arranged at the surface and how exactly it reduces the bare metal work function. In the early days it was assumed that ZrO would form upstanding dipoles with the positively charged zirconium on the outside, thereby counteracting the existing bare tungsten surface dipole (Danielson & Swanson, 1979). Later Tanaka *et al.* (2005) calculated with density functional theory that a more horizontal orientation of the ZrO would be more stable energetically and also has a stronger calculated work function reduction than the vertical configuration.

The reported values for the work function of the Zr/O/W{100} surface show a large spread: from 2.6 eV to 3.3 eV (e.g., Danielson & Swanson, 1979; Samoto *et al.*, 1985; Takeguchi *et al.*, 1991; Lee *et al.*, 1995; Swanson & Schwind, 2008). The large spread is ascribed to both different measurement techniques and different experimental conditions. With respect to the latter it is noted that the work function of a given crystallographic orientation is in itself a function of the contamination level, temperature, and possibly field strength, as these can change the arrangements at the surface. A change in temperature, for example, may change the distribution of ion cores and electrons near the surface due to thermal expansion, atomic rearrangement (roughening, reconstruction), and/or a change of coverage. And besides the static effects, thermal vibrations may also change the charge distribution (Herring & Nichols, 1949). For illustration, in Section 2.3.3 we will briefly address the effect of temperature on the texture of the surface.

The relevant work function is of course that of the Schottky emitter in operating conditions: at elevated temperature and in a field. To determine the work function at these conditions is not straightforward, as will be addressed in Section 3.5.1. Throughout this book slightly different work function values are used.

Generally {100} facets and terraces can be found all over the Schottky emitter surface. Scanning electron microscopy (SEM) images of Schottky emitter cones generally show both smoothly curved sections and faceted areas, which on the cone are often terraced. Figure 2.1 gives a few examples at different length scales. The faceted areas can be identified as {100} and {110} facets.

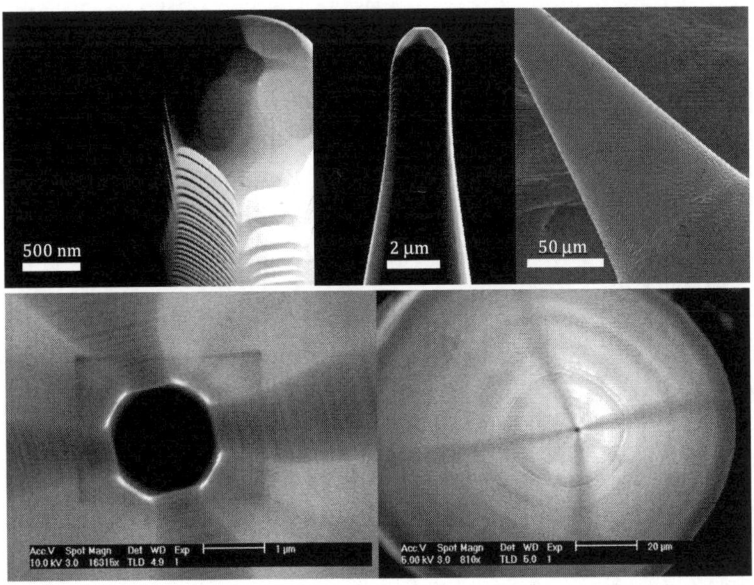

Figure 2.1 SEM images of different Schottky emitters, showing stepped {100} and {110} facets. The bottom series are top-down images. The curved terraces on the two images on the right in the top row, and the four darker bands in the bottom series, indicate {100} facets. The first image shows both {100} and {110} terraces. The {110} terraces are visible in the bottom series as four radial bands, rotated with respect to the {100} bands by 45°.

The effect of the ZrO_x reservoir on the work function of crystallographic orientations other than {100} is much smaller. Emission from a Schottky source is thus predominantly from the {100} facets on the emitter surface. This will be demonstrated in the next section.

2.2 Applying a Bias

To extract electrons from the emitter surface the extractor is biased positively with respect to the emitter. A standard Schottky emitter also always has a suppressor cap surrounding the cylindrical base of the emitter and the heating wire (Fig. i.2). The suppressor is biased negatively with respect to the emitter and thus prevents emission for part of the source. Which part of the source can emit depends on the combination of the suppressor and extractor voltage. The suppressor voltage in commercial applications is usually between –200 V and –800 V with respect to the emitter, the extraction voltage a few kilovolts. Figure 2.2 gives an example of the field along the surface of a Schottky emitter in the standard gun geometry (Fig. 1.2) and a close-up of the typical distribution of equipotential lines near the emitter.

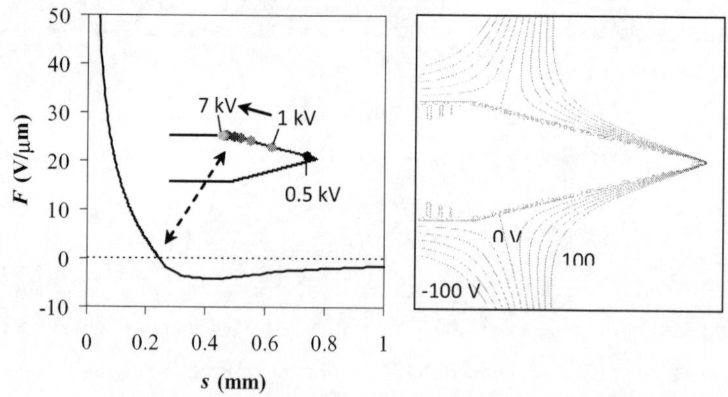

Figure 2.2 Field distribution along the emitter surface for V_{ext} = 5.0 kV, with s the distance along the surface measured from the end facet (left). The inset shows the position of the zero field for different extraction voltages. The right image shows the equipotential lines around the emitter for V_{ext} = 3.0 kV. Both plots apply to the emitter and gun configuration, as in Fig. 1.2, with V_{sup} = –0.3 kV.

Figure 2.2 shows that the field along the surface decreases with increasing distance from the tip and, for the given example, changes direction there where the cone meets the cylindrical part. In this example all of the cone is thus emitting. As shown in the inset on the left the zero-field position is closer to the tip for lower extraction voltages. For constant suppressor voltage, the emitting area thus increases with increasing extraction voltage.

Not only does the emitting area change, but also the "field of view" of the extractor aperture, or the lens effect between the emitter and the extractor. We will demonstrate this below with experimental emission patterns.

An emission pattern can be seen as a distorted and blurred "shadow image" of the emitting areas of a source and can measured with a setup, as that in Fig. 2.3. In the general setup the emission pattern is recorded behind the extractor, where a scintillator (yttrium-aluminum-garnet [YAG] or phosphor) converts electrons into photons. Photos of such setups are given in Fig. 2.4.

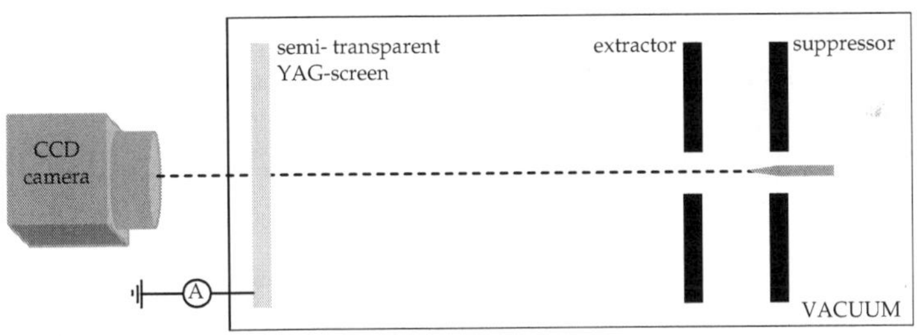

Figure 2.3 Schematic experimental setup for recording emission patterns. Usually the space between the extractor and the screen is kept field free.

For the work described in this book emission patterns have been recorded in two slightly different configurations. The main difference between the setups is the beam energy at the YAG screen. In "setup A" the beam energy was fixed at 7 kV by accelerating the beam between the extractor and the YAG. In "setup B" the beam energy is defined by the extractor potential. The first configuration has the drawback that there is an extraction voltage–dependent lens effect between the extractor and the screen. The second configuration has the drawback that only a limited voltage range can be investigated: for low extraction voltages the photon yield becomes too low for detection.

Figure 2.4 Pictures of the vacuum part of two slightly different experimental setups for recording emission patterns, with the YAG screen on the left and the source part on the right. The inset in the figure on the right shows the typical image as seen by the CCD camera outside the vacuum. *Abbreviation*: CCD, charge-coupled device.

In standard operating conditions most of the total emission from the emitter will not go through the extractor aperture (in commercial guns typically 20 μm or 380 μm in diameter) but will hit the extractor. In that case the emission pattern behind the extractor yields information about only a small part of the emitter: (part of) the end facet. The emission pattern of the end facet will be discussed in more detail in the next chapter.

By decreasing the extraction voltage (and keeping the suppressor voltage constant) the emitting area decreases, but at the same time the "field of view" of the extractor aperture increases. This is illustrated in Figs. 2.5 and 2.6, which give series of different patterns recorded for different extraction voltages at a fixed suppressor voltage. The patterns were recorded for a single emitter at a constant temperature (and for which it can be assumed that the geometrical shape was constant). The patterns are recorded with the setup of Fig. 2.3, with an extractor aperture diameter of 0.38 mm, 0.77 mm between the suppressor and the extractor, and 68 mm between the extractor and the screen (screen diameter 28 mm). To visualize the pattern for low extraction voltages, the beam was accelerated after the extractor to 7 kV (setup A). This gives a (small) voltage-dependent converging lens effect between the extractor and the scintillator. The patterns in Figs. 2.5 and 2.6 are not corrected for this.

From Fig. 2.5 it can be seen that the "field of view" for the lowest extraction voltage is larger than that for the higher extraction voltages: at voltages from 1.5 kV upward the pattern shows only the emission from the end facet, while at the lowest extraction voltage three bright spots are visible, which are associated with emission from the low work function {100} planes on the shanks (Fig. 2.1). It confirms that emission from all crystallographic orientations other than the low work function {100} planes is negligible.

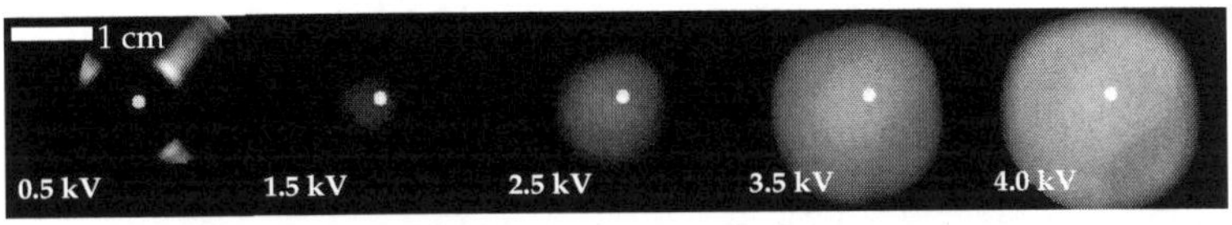

Figure 2.5 A series of emission patterns for different extraction voltages in the experimental setup of Fig. 2.4 (dimensions in text), with V_{sup} = −0.3 kV and an emitter temperature of ~1,800 K. The intensity scales of the images have been adjusted for each pattern. The total current in the patterns was respectively 1.0, 0.6, 2.2, 7.3, and 14.7 μA. N.B. The central bright spot is not caused by electrons but light emitted by the hot source, passing the semitransparent YAG screen.

At the center of the 0.5 kV pattern there still is the end-facet emission spot, but it is very faint and small. Figure 2.6 gives a similar emission pattern from Swanson and Schwind (2008) but with a bright central spot. It is expected that the latter has been recorded at a larger extraction voltage for an extractor with a much larger aperture.

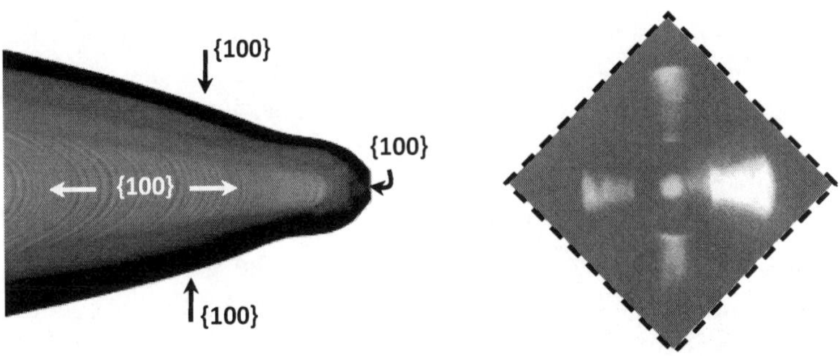

Figure 2.6 SEM image of a Schottky electron source (left) with {100} facets indicated and (right) an emission pattern from Swanson and Schwind (2008), showing emission is from only those {100} facets.

In the 0.5 kV pattern in Fig. 2.5 the fourth lobe is missing due to misalignment. The fourth spot can be made visible by going to even lower voltages, as shown in Fig. 2.7. Note that the emission patterns show texture in the shank emission spots, with higher and lower emission bands, related to the stepped nature of the {100} planes on the cone visible in the SEM images of Fig. 2.1. (Steps in a surface generally have a work function–reducing effect due to the dipole moments associated with the steps (e.g., Besocke *et al.*, 1977; Merrick *et al.*, 2005), but this effect is less dramatic than the effect of the zirconium oxide on the work function of W{100}.)

Figure 2.7 Series of emission patterns for different extraction voltages in the experimental setup of Fig. 2.3 (dimensions in text), with V_{sup} = –0.3 kV and an emitter temperature of ~1,760 K. The intensity scales of the images have been adjusted for each pattern.

Note that the conditions used to record most of the patterns in Figs. 2.5 and 2.7 are not useful at all for any practical applications of the emission, and they are used here only to give insight into the effect of the voltage settings on the emitting area and the lens effect. A temporary change of the field of view of the extractor can be useful to determine the crystallographic orientation of the emitter with respect to its axis. This will be used in Chapter 5.

To support the interpretation of the experimental patterns we reproduce the results qualitatively with simulations. For simplicity we approximate the emitter with a rotationally symmetric shape and launch a few electrons from a few different positions on the surface perpendicular to the surface with the average thermal energy of 1,800 K. Figure 2.8 shows the result for two extraction voltages, a "standard" extraction voltage of 5 kV and a very low extraction voltage of 0.6 kV, both for a suppressor voltage of –0.3 kV with respect to the emitter.

The simulations show that at "normal" operating conditions (left) only the emission from the central end facet passes the extractor aperture (∅ 380 μm) and that the emission from the cone hits the extractor. As shown on the right in Fig. 2.8, at very low extraction voltages the shank emission can protrude the extractor aperture together with the end-facet emission in agreement with the experimental patterns. It can also be seen that for the low extraction voltage the emitting area is indeed reduced: only the electrons close to the tip end can still escape from the emitter, and the emission from parts higher up the cone is now suppressed.

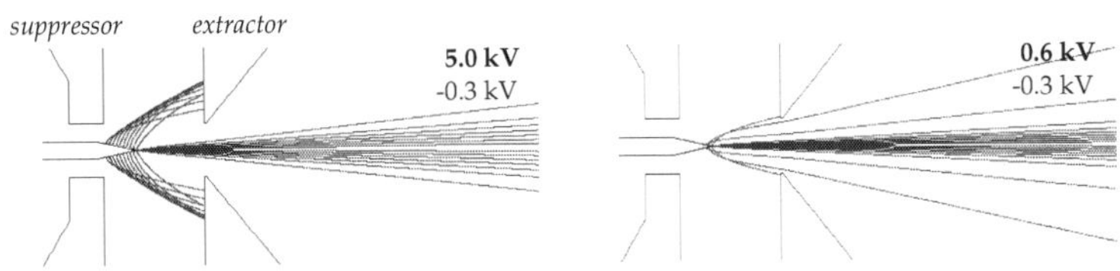

Figure 2.8 Trajectories for electrons launched from a Schottky emitter in the standard gun geometry for two different operating conditions, an extraction voltage of 5.0 kV (left) and 0.6 kV (right), for a constant suppressor voltage of –0.3 kV with respect to the emitter.

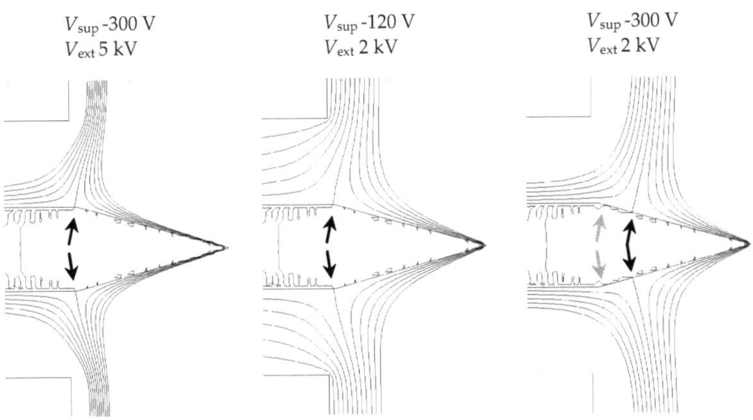

Figure 2.9 The field form remains constant when the suppressor voltage is changed with the extraction voltage proportionally (first two images). The arrows indicate the 0 V line, and the equipotential lines are shown for –100 V to 100 V in steps of 20 V.

In practice usually only one of the voltages is varied: the extraction voltage or the suppressor voltage. Sometimes it can be useful to change the field strength, but keep the emitting area and the field of view of the extractor aperture (the lens effect) constant (see, e.g., Section 3.4). This can be done by varying the suppressor voltage proportionally with the extraction voltage, which ensures the form of the field around the emitter remains constant. This is illustrated in Fig. 2.9, which shows with equipotential lines that for a change in extraction voltage from 5 kV to 2 kV, the field form changes when the suppressor voltage is kept at −300 V but that the form remains constant when the suppressor voltage changes proportionally, from −300 V to −300 × 2/5 = −120 V.

In Section 3.4 we will show that by taking data for a changing field but constant field form can give useful information on the conditions at the emitting surface.

2.3 Applying a Heating Current

The heating current is applied to raise the temperature of the emitter and thereby increase the electron emission. But applying the heating current does not just affect the temperature, as will be shown below.

2.3.1 Temperature

The purpose of applying a heating current is to raise the emitter temperature. The temperature of the surface of a Schottky source is an important parameter for its emission properties. The operating temperature for the source, recommended by the manufacturer, is 1,800 K. At this temperature the emitter is glowing hot, as was already shown in Fig. i.1. The source warms up from resistive heating by running a 2.0–2.5 A current through the V-shaped filament to which the emitter is spot-welded. The actual tip temperature depends on the total of the thermal power generated by Joule heating, thermal conduction, and radiation losses (Dokania *et al.*, 2007).

The manufacturer specifies the heating current that gives the emitter a temperature of 1,800 K for every emitter that is made. This calibration is done in a test setup in which the temperature of the tip is quantified with a thermal imager based on a narrow band pass near an infrared filter. The measured signal is corrected for window transmission and then converted into a temperature using an emissivity value for tungsten. The manufacturer also specifies that near the recommended heating current, the relation between heating current and temperature is linear: 1.25 K/mA.

Figure 2.11 shows, as an example, the relation between the heating current, resistance, and temperature measured for the emitter in Fig. 2.10. The emitter was heated inside a vacuum system, and the temperature was measured through a view port with a disappearing (tungsten) filament pyrometer, pointing at the junction of the filament and the emitter.

Figure 2.10 Glowing Schottky emitter (right) and SEM image of the fixation of the heating filament to the pole (left).

The calibration result from the manufacturer for the emitter in Fig. 2.10 has been indicated in Fig. 2.11 with a dashed line. It can be seen that for temperatures near the recommended operating temperature the measured temperature dependence has approximately the same slope as specified by the manufacturer, but the temperature for the tip determined by the manufacturer is lower. The offset might be ascribed to the temperature difference between the junction and the tip, which can be up to 50 K (McGinn *et al.*, 2004), a systematic error of the pyrometer, or a slightly different environment.

It is noted that in general the environment of a tip in a microscope or lithography machine is slightly different from that in the calibration setup, and the tip temperature at the recommended heating current can thus be slightly different from 1,800 K. For this reason it would be of interest to verify the temperature for the recommended heating current, but in practical applications often the emitter cannot be accessed with a pyrometer.

In theory it should be possible to predict the tip temperature if the relevant dimensions and material properties are known. In that case a heat balance equation can be set up on the basis of the thermal power generated by ohmic heating and liberated through conduction and radiation as in (Dokania *et al.*, 2007). The input parameters for such a balance are the dimensions of the heating filament and the emitter and several material properties such as thermal conductivity, electrical resistivity, and total hemispherical emissivity. Other relevant parameters are the contact of the heating filament to the poles (Fig. 2.10) and the local environment in terms of heat reflective elements such as the suppressor and an extractor. An accurate prediction of the tip temperature is, however, complicated, as the emissivity values of the Schottky emitter materials (heating filaments, emitter, and reservoir) are not known exactly. Also, in practice the length of the heating filament, the size of the reservoir, the contact to the poles, and the local environment can vary, which would require modeling for each emitter and each system separately.

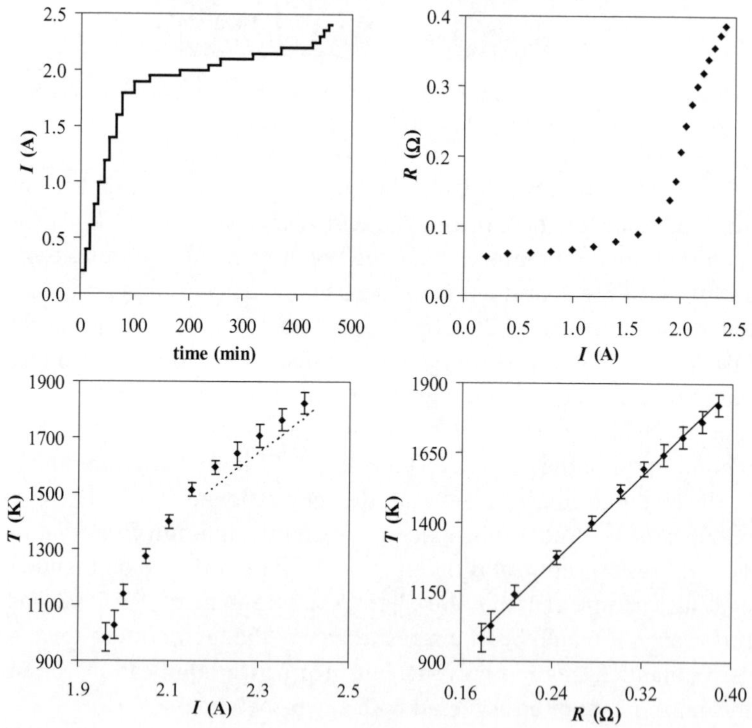

Figure 2.11 Relation between heating current, resistance, and junction temperature for the emitter in Fig. 2.10. Also indicated is the $I(T)$ relation for the emitter tip, as given by the manufacturer (dashed line).

For illustration we give an example of a theoretical temperature prediction. The model geometry is derived from (Dokania *et al.*, 2007). Only one of the two legs of the heating filament is modeled, as it is assumed the two heating legs and the emitter are a symmetric structure. The leg is subdivided into 20 segments. The emitter itself is modeled as additional radiation loss for the segment at the junction, and Joule heating and conduction in the emitter are thus neglected. The only difference with the model in Dokania *et al.* (2007) is that the temperature is not fixed at the end of the leg but at the end of a pole connected to the leg, as in Fig. 2.12. This allows the filament to become warmer, which gives a slightly better representation of the standard Schottky source. We use the same temperature-dependent material properties as used in Dokania *et al.* (2007).

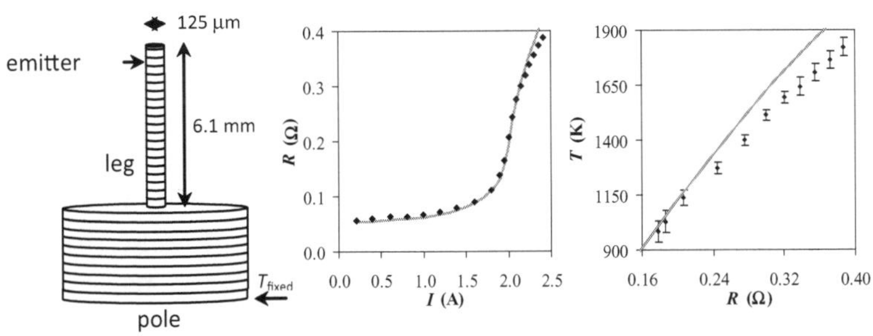

Figure 2.12 Simulation of the relation between heating current, filament resistance, and tip temperature. Black diamonds indicate experimental results and solid lines the simulation results.

It can be seen that the agreement between the model and the experiment is reasonable, except for the relevant section—near the standard operating temperature of 1,800 K. The radiation loss in the model is clearly too low. A possible reason is that the emissivity that is used is too low.

Emissivity is sensitive to surface roughness, adsorbates, native oxides, chemical composition, etc. It is also noted that in the simulation for all materials involved the same emissivity relation is taken. The filament is made of polycrystalline tungsten, the emitter of single-crystalline tungsten, the reservoir of ZrO_x. Matsumoto *et al.* (1999) give a literature overview of the total hemispherical emissivity of tungsten as a function of temperature, showing a considerable spread. Additional data for tungsten are in (Ots05) and Matsumoto *et al.* (2000). And from the literature it seems not unlikely that the reservoir has a higher different emissivity than the emitter itself (Wade, 1959; Jain, 1970; Ferriere *et al.*, 2000). It is noted that the agreement between the experiment and the simulation can be improved considerably if the tip area or the emissivity is increased (not shown).

The above suggests it may be possible to estimate the tip temperature from simulations and a measured filament resistance with reasonably accuracy, but for definite conclusions a more thorough investigation is required. In this book, to determine the temperature of an emitter, we will use the relation between heating filament and temperature as calibrated by the manufacturer.

2.3.2 Tip Protrusion (Field Strength)

By heating up the emitter, the emitter will expand. This changes the protrusion of the emitter from the suppressor cap and thus the field distribution across the emitter surface. To illustrate the effect of the heating current on the tip protrusion, a Schottky emitter was heated by resistive heating inside an SEM chamber, as shown in Fig. 2.13. Figure 2.14 gives the measured relation between heating current, filament resistance, and protrusion. The bulk of the emitter and the wire is tungsten, which has a thermal expansion coefficient of ~6 × 10^{-6} K^{-1} in the appropriate temperature range (Wang & Reeber, 1998). With this a rough estimate of the change in protrusion between room temperature and operating temperature can be made. Assuming a total length of ~4 mm (the emitter plus part of the filament) heats up to 1,800 K, the expansion can be expected to be ~36 μm.

Figure 2.13 A CCD image of a Schottky emitter with a suppressor cap inside an SEM chamber (left) and an SEM image of the emitter cone protruding from the suppressor cap (right).

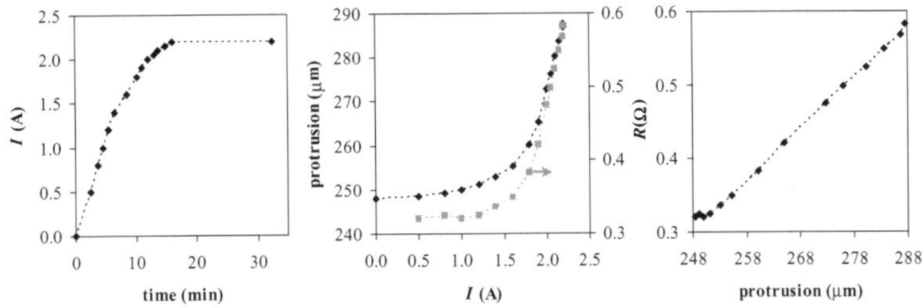

Figure 2.14 Effect of the heating current on tip protrusion and filament resistance.

A different tip protrusion gives a different field at the emitter. This is illustrated in Fig. 2.15, which gives the field strength at the center of the end facet as a function of the protrusion, for an emitter in the standard gun geometry, operated at a suppressor voltage of -0.3 kV and an extraction voltage of 6 kV with respect to the emitter.

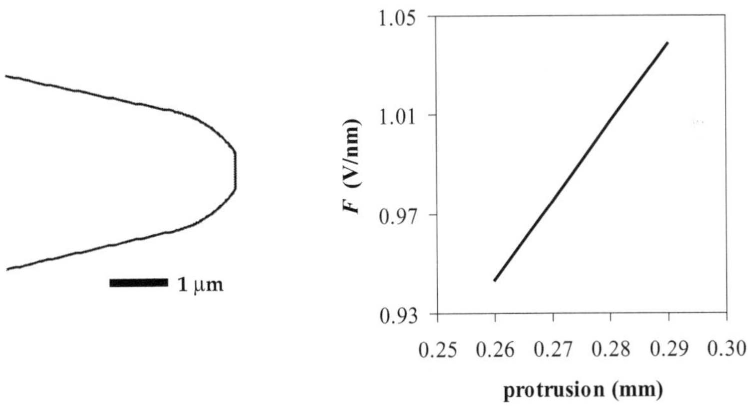

Figure 2.15 Effect of a change in protrusion on the field strength at the center of the end facet for an emitter with the given tipend geometry (left) in the standard gun geometry (0.76 mm between extractor and suppressor), operated for a suppressor voltage of -0.3 kV and an extraction voltage of 6 kV with respect to the emitter.

In the example the difference in field strength for a 30 μm extra protrusion is 10%. Because the emission is a strong function of the field strength the change in protrusion upon heating up the source is something to take into account for more accurate predictions.

2.3.3 Surface Properties (Work Function)

Finally, by heating up the source, the work function of the surface might change. In general many surface properties, such as the equilibrium adsorbate or defect concentration or the structure of the surface, can be temperature dependent, and this can affect the work function.

For example, for the Zr/O/W{100} surface there are indications that the structure at room temperature is different from that at operating temperature (Tamura *et al.*, 2005; Lee *et al.*, 1995; Irokawa *et al.*, 1996).

The temperature dependence of the work function of the Zr/O/W{100} surface will be addressed in more detail in Section 3.5.1. Here, merely for illustration, we show the effect on the texture of the surface of a Schottky emitter upon applying a heating current. Both the Schottky emitter and the conditions are not "standard," and it is thus not expected to be an exact representation of what happens for a normal, emitting source. The tip of the emitter that was used had been arced in a prior experiment and was reshaped by etching (in NaOH) right before the experiment. And the emitter was heated inside an SEM chamber with a pressure of the order of 1×10^{-6} mbar, which is a factor 1,000 higher than in the standard operating conditions.

Figures 2.16 and 2.17 give the changes for two different areas on the surface. The SEM picture series in Fig. 2.16 shows an area with a small pyramidal shape on the surface. Figure 2.17 shows the evolution for a vicinal area. It can be seen in Fig. 2.16 that at a 2.05 A heating current, for which the tip had just started glowing, the surface is already changing. The picture taken at a heating current of 2.15 A shows a highly (micro)faceted and ordered structure, which is lost to some extent over time, as judged from the next image. It is noted that similar hill-and-valley faceting has been observed for oxygen-covered tungsten in the temperature ranges 860–1,300 K and 1,640–1,800 K (Szczepkowicz & Bryl, 2004). By operating at 2.20 A the microfaceting on the pyramid is lost completely, and after a night at 2.20 A, and cooled to room temperature, the pyramid itself had gone, leaving a smooth surface (not shown).

In conclusion, applying the heating current affects not only the tip temperature but also the local field strength and possibly the effective work function.

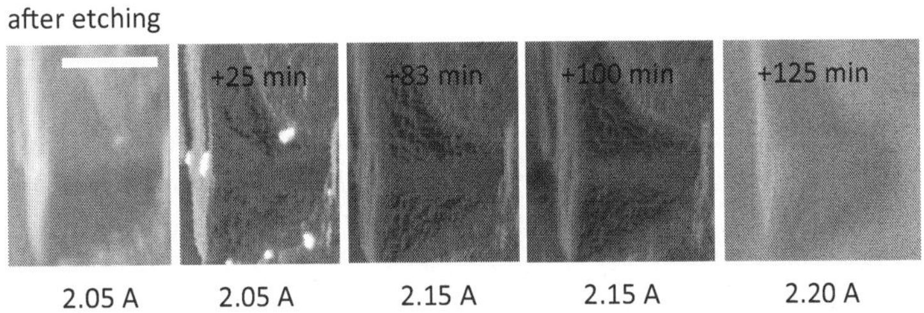

Figure 2.16 Heating an emitter inside the SEM chamber: the evolution of a pyramidal shape on the tip of a Schottky source that first had been etched in NaOH. The time indicated is the time that has elapsed with respect to the first image on the left. The scale in the first image is 1 μm.

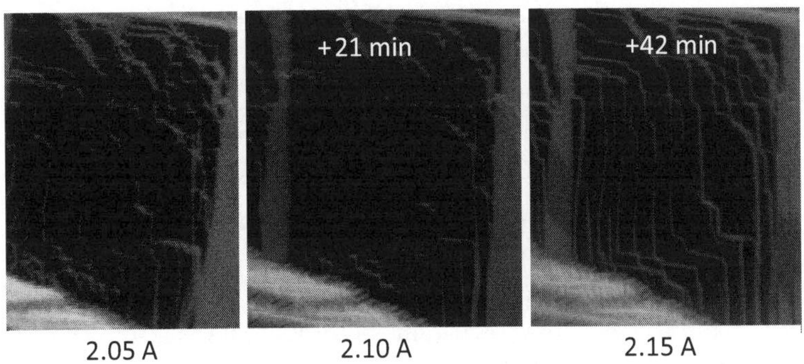

Figure 2.17 Heating an emitter inside the SEM: the evolution of a vicinal area on the tip of a Schottky emitter that first had been etched in NaOH. The number of steps in a surface can change the effective work function. The scale bar in the left image is 500 nm.

2.4 Total Emission Current

With the field distributions and work function variations across the surface and the emitter temperature we can now estimate the total current emitted by a Schottky source with the theory of Chapter 1. It is noted that because the {100} area can vary considerably from emitter to emitter, it would require careful inspection of the emitter surface for anything better than a rough estimate. For the rough estimate we would like to make now we assume the conical part of an emitter is an octagonal pyramid of height 250 μm and base

125 μm with alternating faces of stepped {100} planes and stepped {110} planes. Any emission from other than the {100} planes is neglected. For this geometry the total {100} area on the cone can be estimated with $4C_f \frac{1}{2} bH$, where C_f is a correction to take into account that due to the stepped nature not all of the area has a low work function. On the basis of SEM images we take for α_f 10°, and we use a correction factor C_f of ½ (Fig. 2.18).

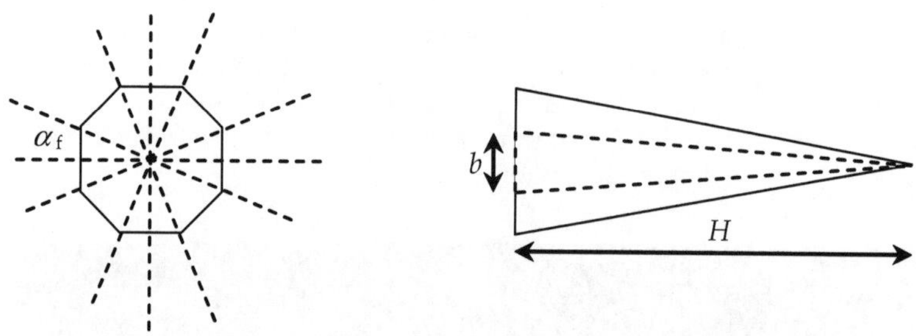

Figure 2.18 Geometry for estimating the total emission: top-down (left) and side (right) views.

Note that which part of the available {100} area will actually contribute to the emission depends on the voltage settings.

Figure 2.19 gives the estimated total emission as a function of extraction voltage, for a constant suppressor voltage of −0.3 kV with respect to the emitter, in the standard gun geometry, a work function of 3.05 eV, and a temperature of 1,800 K. For comparison an experimental result is also given.

The deviation between calculation and experiment is considerable, but the results are in the same ballpark. Because the current density is such a strong function of temperature, field, and work function, and the approximation of the emitting area is rather rough, the agreement is considered reasonable.

The conclusions of this chapter are as follows. Emission from the Schottky emitter surface other than with a {100} orientation is negligible. In the standard gun geometry with standard operating conditions the suppressor suppresses the emission from the emitter up to approximately the point where the cylindrical base meets the cone. The total emission from a standard Schottky emitter at the standard operating temperature can be up to a few hundreds of microamperes. In normal conditions most electrons are collected at the extractor, and only the facet emission can protrude a standard

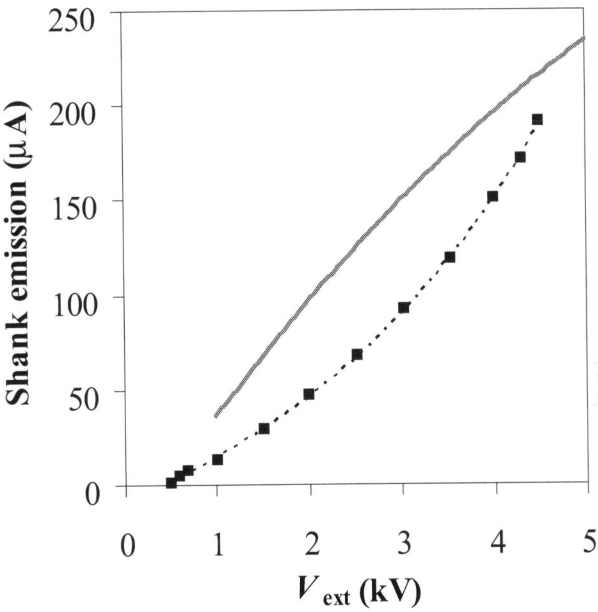

Figure 2.19 Calculated (gray line) and measured (black diamonds) cone emission as a function of the extraction voltage for a standard Schottky emitter at 1,800 K in the standard gun geometry with a constant suppressor voltage of –300 V. The work function for the calculation was taken to be 3.05 eV.

380 μm extractor aperture at a ~0.5 mm distance from the tip. We will look at this facet emission in more detail in the next chapter.

Chapter 3

Emission from the End Facet

Of all electrons emitted by a Schottky source, the emission from the end facet is the most important because it delivers electrons that are used for probes and beams in commercial applications. Up till recently focus has been on the emission properties of the center of the facet. This is because most commercial systems use only a very small percentage of the total current emitted by the facet (typical beam currents of the order of picoamperes to nanoamperes compared to typical facet currents of tens of microamperes). With the recent efforts to develop multibeam systems (Zhang & Kruit, 2007; Young *et al.*, 2009) the interest in the emission properties of all of the facet has increased and also its stability in terms of its shape.

The effect of the operating conditions on the uniformity of the emission properties across the facet, and on the stability of the shape of the facet, can be investigated by looking at the facet emission pattern. The emission pattern is a strong function of the conditions. It can look quite different for a single emitter, even for a constant geometrical shape of the emitter. This was already shown in Fig. 2.5. For correct interpretation of the emission pattern we need to know the translation between the current density distribution on the facet and the current density distribution in the emission pattern. This will be the focus of this chapter. In Chapter 5 the emission pattern will be used to monitor changes in the physical shape of the emitter.

The emission pattern is a current density distribution that is measured in a plane just behind the extractor. The current density distribution across the *facet* can be calculated with the electron

Physics of Schottky Electron Sources: Theory and Optimum Operation
Merijn Bronsgeest

ISBN 978-981-4364-79-9 (Hardcover), 978-981-4364-80-5 (eBook)
www.panstanford.com

emission theory of Chapter 1 for a given temperature and work function if the field distribution across the facet is known. The latter can be found with the charge density method. But how does the current density distribution on the facet translate into an emission pattern?

We will first investigate the lens effect between the facet and the extractor. On the basis of this investigation we will make the approximation that all facet emission can be traced back to a single "point source" on the optical axis. In that case the emission pattern can also be considered an angular intensity distribution, a distribution in the angular current density, or current per solid angle, emitted by a point source. In this chapter we will assume that all electrons have zero energy upon emission. In practice this is not the case (see Chapter 1), but the effect of this on the shape of the emission pattern is only small. We will discuss the effect of the energy distribution of the electrons in more detail in Chapter 4.

With the results from the lens effect analysis, the current density on the facet can be converted into an angular intensity distribution, which can be compared with experimental emission patterns. In Section 3.1 we investigate the conversion of a current density distribution to an emission pattern for one specific case. In Section 3.2 we will discuss the effect of the voltage settings (in more detail than we did in the previous chapter) and in Section 3.3 the effect of the emitter geometry. Section 3.4 is addressed to "Schottky plots," an experimental method that can yield the conversion factor between the voltage applied and the field at the emitting surface. In Section 3.5 the effect of the emitter temperature is discussed.

3.1 The Facet Extractor Lens

3.1.1 At the Facet

To find the emission from the end facet it is first of all required to know the field distribution across the facet. This can be done with the charge density method, which has already been introduced in the previous chapters. For the following example (Figs. 3.2–3.5) we take the emitter geometry as given in Fig. 3.1, and the gun geometry from Fig. 1.2, with a protrusion of 242 nm and a distance between the suppressor and the extractor of 0.75 mm. The field distribution across the facet is given in Fig. 3.2.

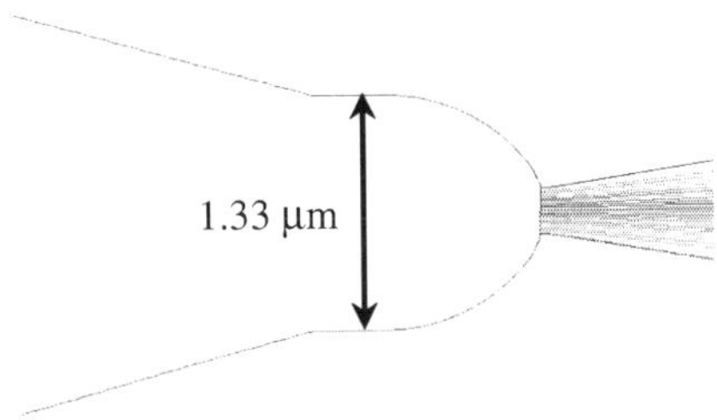

Figure 3.1 Emitter geometry used for the investigation of the lens effect between emitter and extractor. The facet diameter is 286 nm, and the angle at the facet is 26°.

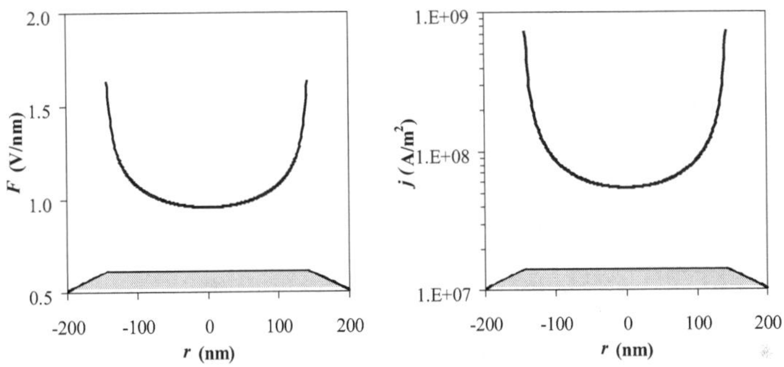

Figure 3.2 Distribution of electric field and current density across the facet for an extraction voltage of 5 kV and a suppressor voltage of −0.3 kV with respect to the emitter. T = 1,800 K; ϕ = 2.93 eV.

As can be expected, Fig. 3.2 shows that the field is stronger toward the facet edge. With the theory of Chapter 1, the field distribution can be converted into a current density distribution for a given temperature and work function. We assume a temperature of 1,800 K and a work function 2.93 eV, all constant across the facet. This result is also shown in Fig. 3.2: there is enhanced current density near the facet edge, similar as for the field distribution.

Before we will start using results as in the left image of Fig. 3.2 more extensively, it is considered useful to become more specific on the details of the simulations:

"The field distribution across the facet," as in Fig. 3.2, actually is the field strength evaluated at specified points in a plane close to the position of the charge segments that constitute the end facet on the emitter contour. Parameters that can affect the result of the calculation are the number of "segments" on the end facet, the position of the plane for which the field is evaluated with respect to the plane of the charge segments, the requested accuracy, and the exact geometry near the facet edge. The exact shape of the emitter in the vicinity of the facet edge, while in operation, is not known.

For the position of the evaluation plane with respect to the plane of the facet we use at least two times the segment length (as recommended by software designers; http://www.electronoptics.com). If the field evaluation plane is very close to the plane of the end facet electrode, the field distribution across the facet will display small oscillations, with a period that matches the segment length on the facet electrode. For most calculations the segment length on the facet was 1 nm, and the field is thus evaluated at a distance of 2 nm.

3.1.2 Lens Properties

To investigate the lens effect between the facet and the extractor, we trace rays that are launched from different positions on the facet and analyze their position and angle in a test plane in the field-free zone behind the extractor, at 2.5 mm from the emitting facet. Figure 3.3 shows the relation between the initial launch position on the facet, and the final angle with the optical axis at a test plane in the field-free zone, for the given voltage settings.

If the equipotential lines between the facet and the field-free zone would form a perfect electrostatic lens, the curve in Fig. 3.3 would be a straight line. This is not the case: the rays launched further away from the facet center experience a stronger total angular deflection relative to the rays launched from positions closer to the facet center. This is consistent with a negative lens effect with positive spherical aberration.

By extrapolating the tangents of the rays back from the field-free zone toward the emitter we find the position where they cross the optical axis. The position of this point with respect to the position of the facet is given in Fig. 3.4 as a function of the launch position.

Figure 3.4 shows that the trajectories seem to come from behind the facet (at ~30 μm) but do not all originate from exactly the same point: the position of the axis crossing is further behind the facet for rays launched further away from the facet center. It is noted that a single negative lens with positive spherical aberration would show the opposite effect and that the facet extractor lens thus might be better described with a lens *system*.

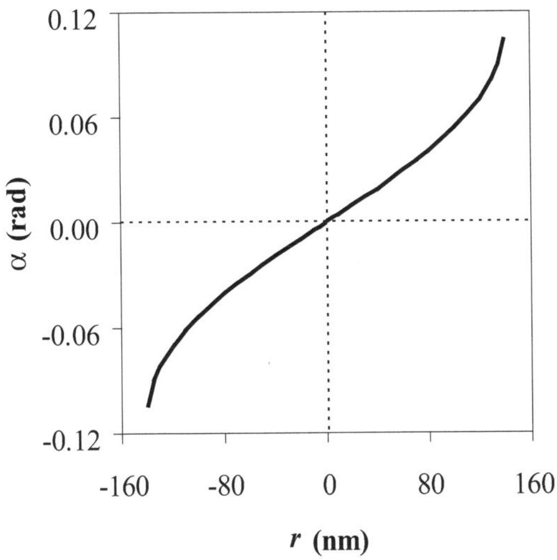

Figure 3.3 Final angle of the ray at a test plane in the field-free zone behind the extractor as a function of the initial launch position on the facet. V_{sup} = −0.3 kV and V_{ext} = 5.0 kV with respect to the emitter.

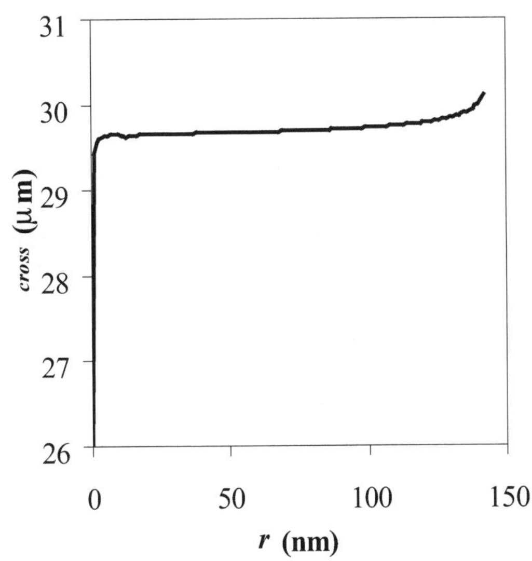

Figure 3.4 Relation between the launch position on the facet and the point where the tangent to the trajectory in the field-free zone crosses the optical axis. z_{cross} is the distance behind the facet, inside the emitter. V_{sup} = −0.3 kV and V_{ext} = 5.0 kV with respect to the emitter.

If we analyze the trajectory of a single electron in more detail (Fig. 3.5) we see that close to the facet the ray is deflected outward, consistent with a negative lens, followed by a section in which the beam divergence is reduced toward the extractor due to the acceleration, and finally, the extractor aperture itself provides again a negative lens effect.

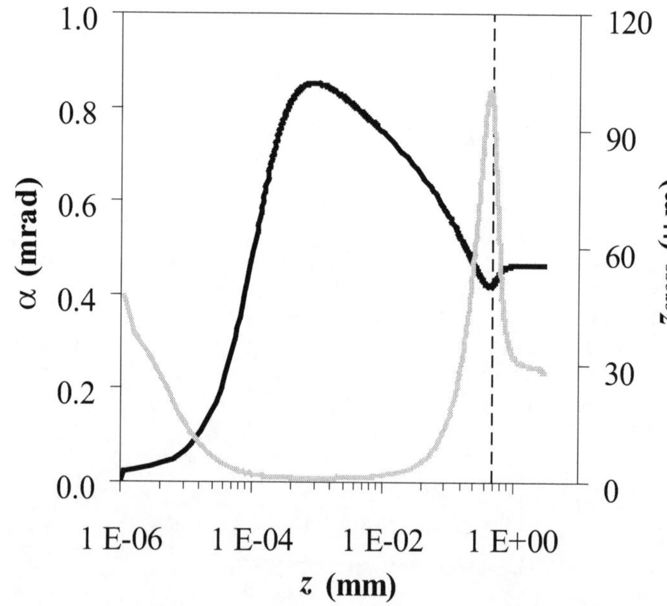

Figure 3.5 Ray angle as a function of the distance along the optical axis for an electron launched from the facet at 1 nm from the optical axis. $V_{sup} = -0.3$ kV and $V_{ext} = 5.0$ kV with respect to the emitter. The gray curve indicates the distance between the facet and the position where the extrapolated tangent to the trajectory crosses the optical axis. The dashed line indicates the position of the extractor.

In conclusion, the total negative lens effect is the result of two negative lenses and the converging effect of the acceleration. To explain that the final angular deflection is stronger for rays launched further away from the optical axis, but that at the same time the point on the optical axis where these outer rays appear to be originating from is behind the point where the inner rays appear to be coming from, one of the lenses has to have positive spherical aberration, and one has to have negative spherical aberration.

For the calculation of the emission pattern we can safely ignore the small variation of the axis-crossing position with the final trajectory angle and assume all electrons originate from the same point, but we will come back to this when we discuss the virtual source in Chapter 4.

As for the evaluation of the field distribution across the facet, the ray-tracing result is sensitive to the number of charge segments on the end facet, the exact shape of the facet and the area just behind the facet, and the position of the plane from which electrons are launched. It is also noted that for accurate ray tracing it is important to make small steps near the tip, where the deflection is strong. Similar as for the evaluation of the field at the facet, electrons are launched at a distance of at least two times the segment length (as recommended by the software designers, http://www.electronoptics.com), to avoid the small oscillations in the potential due to the finite-size facet segments. The initial velocities are zero, except for the velocity parallel to the optical axis, which is set at 100 m/s. The step size near the emitting surface is 1 nm.

3.1.3 Behind the Extractor

3.1.3.1 The angular intensity of the source

The current density on the emitting surface, for which we have the theoretical expressions, cannot be measured. What can be measured relatively easy is the facet emission pattern or part of it. In practice often the central part of the facet emission pattern is probed by measuring the current that comes through a small aperture. This is generally done to characterize the source, and the measured current is then translated into an angular current density

$$I' = \frac{I}{\Omega}, \tag{3.1}$$

with the measured current I and a solid angle Ω. The solid angle Ω is the area of a spherical cap that is cut from a unit sphere by a cone with the half opening angle α

$$\Omega = 2\pi(1 - \cos\alpha), \tag{3.2}$$

with α determined from the geometry of the setup

$$\tan\alpha = \frac{d}{2l}, \tag{3.3}$$

with d the aperture diameter and l the distance between the center of the sphere (the position of the "point source"; Fig. 3.6) and the aperture. Generally, l can be replaced by the distance between the emitter and the aperture because the distance between the tip end and the point where the electrons seem to originate from (the "point source," at ~30 μm from the tip end) is much smaller than the distance between the tip end and the aperture.

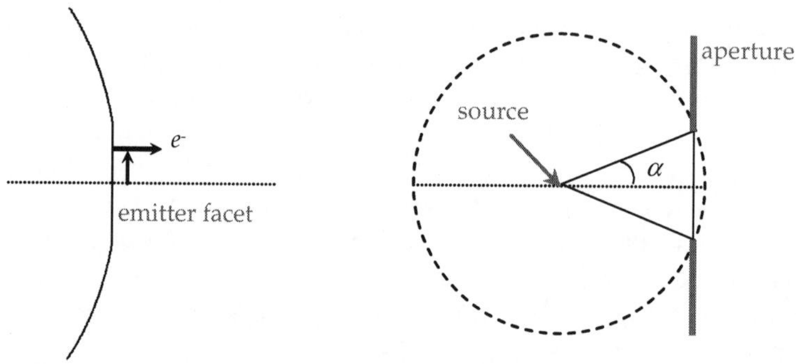

Figure 3.6 Illustration to clarify the relation between theoretical current density and the angular intensity measured with an aperture.

The aperture is generally positioned on the optical axis, and the calculated angular current density is then taken to be the "angular intensity of the source." The angular intensity of a source can often be measured for emitters in commercial systems (see appendix A). Usually the acceptance angle of the aperture that is used is small, and the solid angle is approximated with

$$\Omega = 2\pi(1-\cos\alpha) = 4\pi\cdot\frac{1}{2}(1-\cos\alpha) = 4\pi\sin^2\frac{\alpha}{2} \approx 4\pi\frac{\alpha^2}{2^2} = \pi\alpha^2 . \quad (3.4)$$

Now, to connect the experimental angular intensity of the source and the current density at the emitting facet, note that the measured current I should correspond to $j\pi r^2$, with j the current density in the facet center, and the facet area πr^2 that delivers the electrons that reach the Faraday cup. The relation between current density and angular current density for a small aperture can thus be written as

$$I' = \frac{I}{\Omega} = \frac{j\pi r^2}{\pi\alpha^2} = \left(\frac{r}{\alpha}\right)^2 j . \quad (3.5)$$

The ratio r/α captures the lens effect experienced by the beamlet leaving the facet center and is equal to the inverse of the slope of the curve in Fig. 3.3 at r = 0 nm. In Fig. 3.3 it is 2.12 μm/rad, which for a work function of 2.93 eV and a temperature of 1,800 K would give an angular intensity of 0.25 mA/sr.

The relation between the current density on the facet, and the angular intensity of the source, has been investigated in other studies, yielding different terminology to indicate the proportionality factor between I' and j. Fujita and Shimoyama (2005a, 2005b, 2006) use f^2 and name f the electron gun focal length. Swanson and Schwind (2008), in *Handbook of Charged Particle Optics*, write the proportionality between I' and j as

$$I' = \left(\frac{r}{\alpha}\right)^2 j = \left(\frac{\varphi r_{\text{tip}}}{\alpha}\right)^2 j = \left(\frac{r_{\text{tip}}}{m_\alpha}\right)^2 j, \tag{3.6}$$

in which r_{tip} is the tip radius and m_α is the so-called angular magnification:

$$m_\alpha = \frac{\alpha}{\alpha_{\text{tip}}}, \tag{3.7}$$

where α_{tip} is the "internal" angle given by the ratio of the distance of the launch position from the optical axis, and the tip radius. The definition of that internal angle is shown in Fig. 3.7.

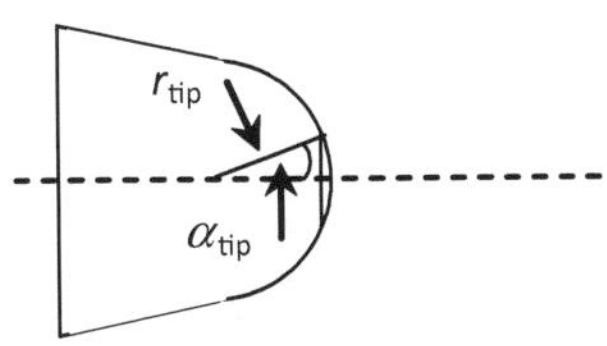

Figure 3.7 Definition of α_{tip} in Eq. (3.7).

The relation between the facet launch position r and the ray angle α is a strong function of the emitter and gun geometry and the applied potentials. This will be discussed in Sections 3.2 and 3.3.

3.1.3.2 The full facet emission pattern

By moving the aperture in the plane perpendicular to the optical axis, just behind the extractor, measuring at each position the current through the aperture, one would build up the full facet emission pattern. To connect the full facet emission pattern to the associated current density distribution across the facet would in principle require a calculation of the solid angle for each aperture position or, if the facet emission pattern is required with a charge-coupled device (CCD) camera, for each pixel. However, if the facet beam is rotationally symmetric, the translation between the angular current density distribution behind the extractor and the current density distribution on the facet is easier because one can make use of the solid angle of annular apertures.

For an annular aperture the solid angle is

$$\Omega = 2\pi \sin\alpha d\alpha. \tag{3.8}$$

This gives the relation between current density and angular current density for an annular aperture (see also Fig. 3.8):

$$I' = \frac{I}{\Omega} = \frac{j2\pi r dr}{2\pi \sin\alpha d\alpha} = \left(\frac{r dr}{\sin\alpha d\alpha}\right) j, \tag{3.9}$$

with r is the distance between launch position and optical axis, α the final ray angle of the launched electron in the field-free zone behind the extractor, and $dr/d\alpha$ the derivative of the relation between the launch position and the final angle.

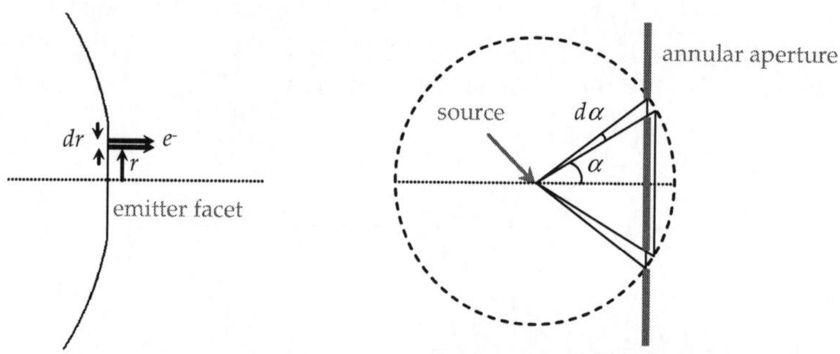

Figure 3.8 Illustration to clarify the relation between theoretical current density and the angular intensity for an annular aperture.

With Eqs. (3.5) and (3.9) one can construct the full facet emission pattern. As a first example we give in Fig. 3.9 the emission pattern that corresponds to the facet current density distribution in Fig. 3.2. Note that the shape of the distribution has changed completely going from the facet to behind the extractor, as a result of the spherical aberration of the facet extractor lens.

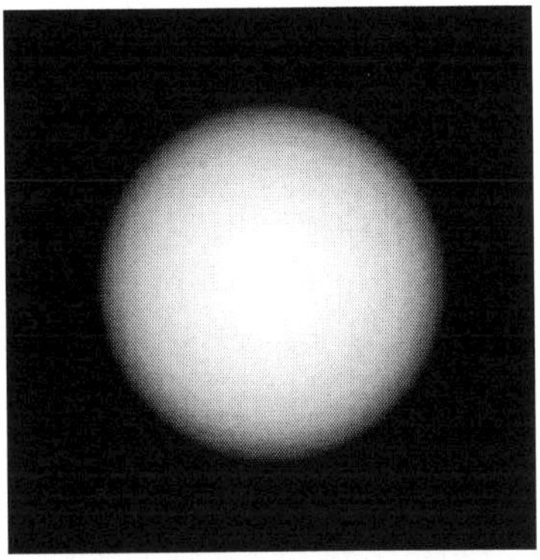
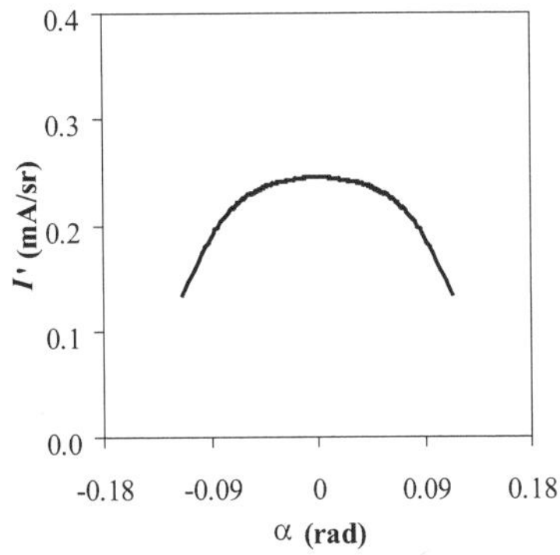

Figure 3.9 Simulated emission pattern from the current density distribution in Fig. 3.2 and the ray-tracing results in Fig. 3.3.

The exact shape of the full facet emission pattern strongly depends on the voltage settings, the emitter geometry, and the emitter temperature. This will be investigated below.

3.2 The Effect of the Voltage Settings

3.2.1 Different Options

The voltage settings for the suppressor and extractor determine the field strength at the facet and the lens effect experienced by the facet emission. In practice the voltage settings are changed to change the emission properties and/or the beam energy at the extractor.

One has the following options:

- Change the extraction voltage at a constant suppressor voltage; in this case the field, field form, and the beam energy at the extractor will all change.
- Change the suppressor voltage at a constant extraction voltage; the field strength and field form will change, but the beam energy at the extractor is constant. This can be useful if the gun design is optimized for a fixed beam energy at the extractor.
- Change both the suppressor and the extractor by the same amount; the field strength and field form remain constant, but the beam energy at the extractor will change. This can be

useful if the emission properties are as desired but the beam energy is not optimized for best gun performance yet.

- Change the suppressor and the extractor voltage in proportional amounts; the beam energy and field strength will change, but the field form is constant. This can be useful to determine how an applied voltage translates to a field strength at the emitter surface (see Section 3.4).

As an example, Fig. 3.10 shows a series of experimental emission patterns for constant suppressor voltage and varying extraction voltage, for constant extraction voltage and varying suppressor voltage, and for extractor and suppressor voltages changed by the same amount. The gun geometry is the standard gun geometry (e.g., Fig. 1.2) with a distance between the suppressor and the extractor of 0.76 mm and a room-temperature protrusion of 249 μm. The actual shape of the emitter at the time of the measurement is not known exactly, because the data are taken after a couple of weeks of continuous operation, but a scanning electron microscopy (SEM) image of the emitter before the experiments is given in Fig. 3.11.

It is noted that for the final series in Fig. 3.10, for which the suppressor and the extractor are changed by equal amounts, the patterns are very close, as expected, but not exactly the same: the pattern becomes slightly smaller going from left to right. It turns out that the field strength and field form are not a 100% equivalent for the different voltage settings. Simulations show that there is a small field strength difference for the settings on the left end of the bottom series of Fig. 3.10 compared to that at the right end (4%), and there is indeed a difference in current measured in the patterns of 13%. And ray tracing yields that the lens effect that is experienced near the emitter is equivalent, but the ray deflection near the extractor aperture is slightly different for the different conditions, with a more compressed emission pattern for the higher extraction voltage, also in agreement with the experimental series.

3.2.2 The Effect of Changing the Extraction Voltage

In the remainder of this section we concentrate on the effect on the 1,800 K emission pattern and angular intensity of the source of a varying extraction voltage and a constant suppressor voltage.

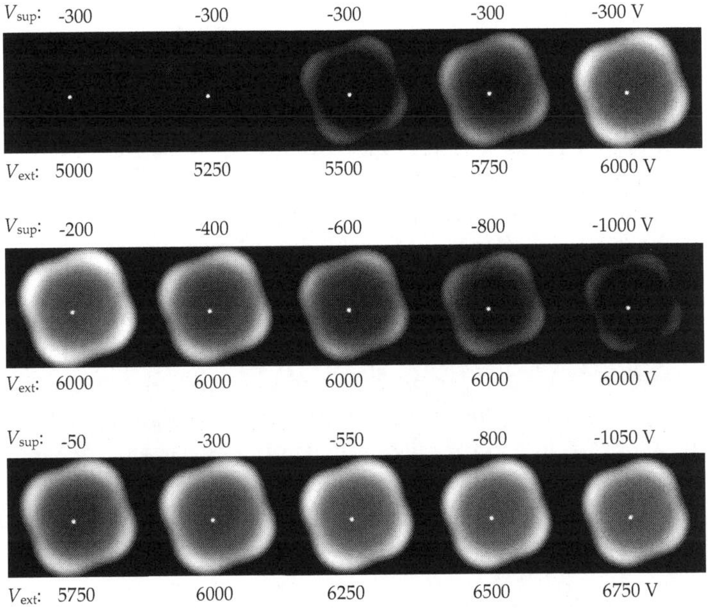

Figure 3.10 Effect of different voltage settings on the emission pattern. (Top) Changing field, field form, and beam energy at the extractor. (Center) Changing field and field form but constant energy at the extractor. (Bottom) Constant field and field form but changing beam energy at the extractor. The patterns are measured for the emitter in Fig. 3.11, at ~1,750 K in the standard gun geometry, without changing the camera settings.

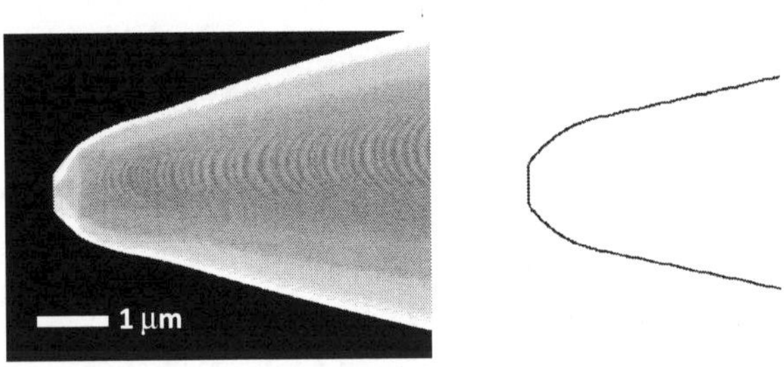

Figure 3.11 SEM image of a Schottky emitter and the contour of a rotationally symmetric emitter derived from the SEM image that will be used for simulations.

The experimental data in this section are taken for the emitter shown in Fig. 3.11, operated in the standard gun geometry (Fig. 1.2), with a suppressor–extractor distance of 0.76 mm, a room-temperature protrusion of 249 μm, and a constant suppressor voltage of −0.3 kV (unless stated otherwise) with respect to the emitter. For the simulations we approximate the emitter with the rotationally symmetric shape on the right in Fig. 3.11, and we add 30 μm to the room-temperature protrusion to take into account thermal expansion upon heating the emitter to 1,800 K. For the work function we use 3.04 eV.

First the full emission pattern will be discussed, and then we zoom in onto the center of the emission pattern and address the effects on the angular intensity of the source.

Figure 3.12 gives the field distributions across the facet of the Schottky emitter for different extraction voltages. The associated current density distributions for 1,800 K are also given. It can be seen that the field and the current density increase with increasing extraction voltage, and for all settings the field and the current density are stronger at the facet edge with respect to that at the facet center.

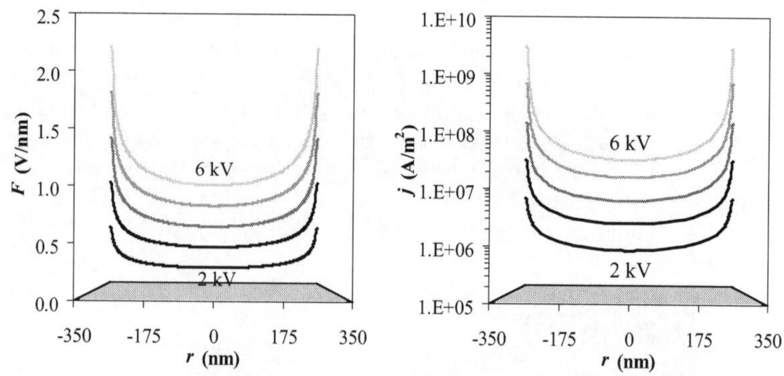

Figure 3.12 Distribution of the field (left) and the current density (right) across the facet for different extraction voltages.

3.2.2.1 From the facet toward the extractor

The consequences of changes in the distribution at the facet can be translated to changes in the distribution behind the extractor using the lens effect. The lens effect as a function of the extraction voltage is given in Fig. 3.13, together with the calculated angular current density distributions. It is noted that the variation in the lens effect in Fig. 3.13 is hardly visible, but it will be shown later that it is not negligible.

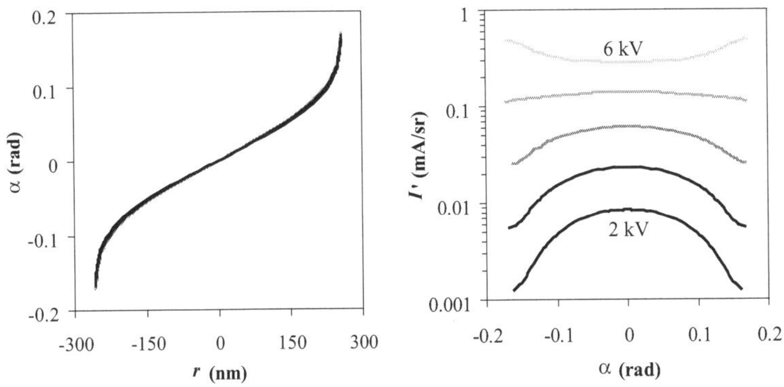

Figure 3.13 Lens effect (left) and angular intensity distributions for different extraction voltages.

With respect to the angular intensity distributions in Fig. 3.13 it is noticed that while the current density distributions on the facet (Fig. 3.12) all showed enhanced intensity toward the facet edge, the (angular) current density distributions behind the extractor show a transition from a bell-shaped to a fairly flat profile to a profile with increased intensity at the edges. This is in agreement with experimental results as given in Tuggle (1984) and Swanson and Schwind (2008) and will also be shown below.

The voltage range for which a pattern can be recorded without changing the camera settings is limited. Figure 3.14 compares simulated emission patterns with experimental patterns taken with constant camera settings for an extraction voltage of 5.25–6.25 kV. Figure 3.15 compares cross sections of the patterns. Figures 3.16 and 3.17 do the same for an extraction voltage of 4.25–5.25 kV.

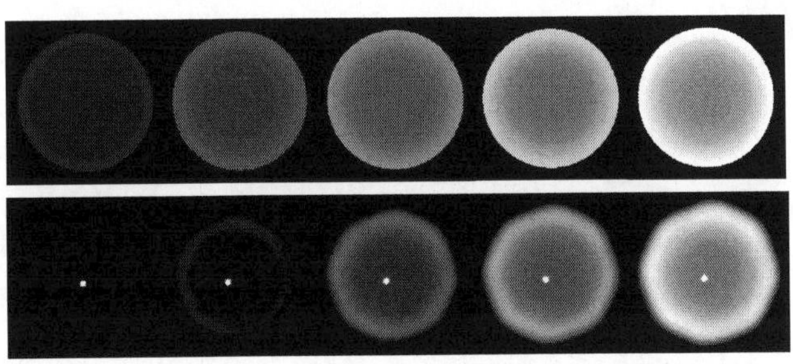

Figure 3.14 Experimental (top) and simulated (bottom) emission patterns for an extraction voltage of 5.25 kV to 6.25 kV from left to right in steps of 0.25 kV.

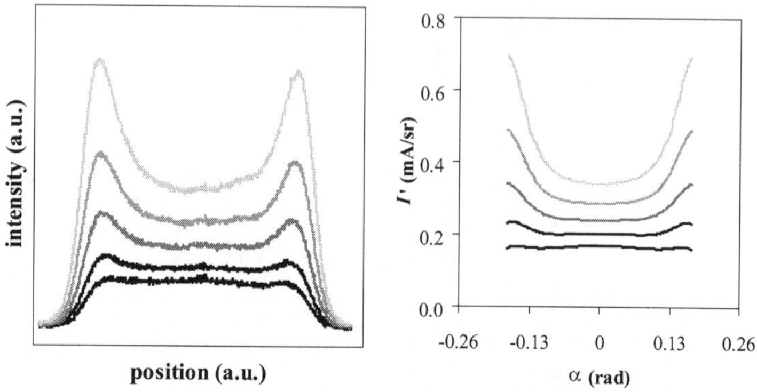

Figure 3.15 Experimental and simulated intensity distributions associated with Fig. 3.14.

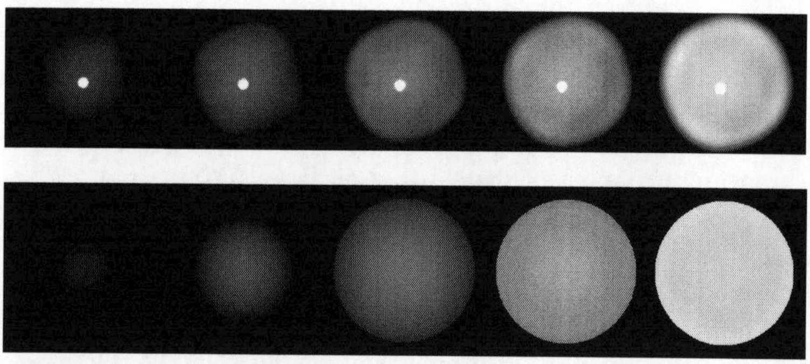

Figure 3.16 Experimental (top) and simulated (bottom) emission patterns for an extraction voltage of 4.25 kV to 5.25 kV from left to right in steps of 0.25 kV.

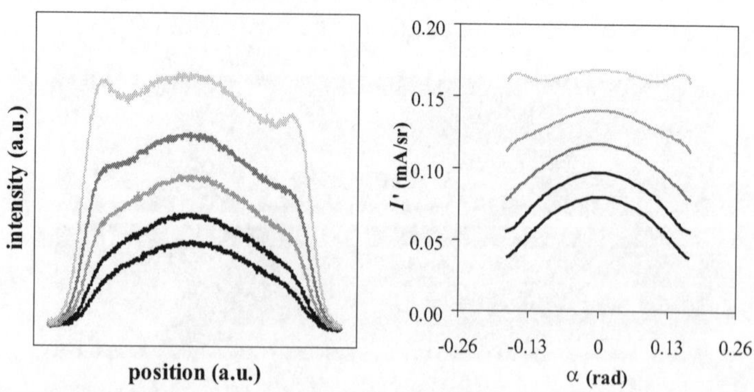

Figure 3.17 Experimental and simulated intensity distributions associated with Fig. 3.16.

The agreement between the experimental and simulation results is good qualitatively but also reasonable quantitatively. This is concluded from Figs. 3.18 and 3.19.

Figure 3.18 compares the measured and calculated facet currents. Two experimental curves are shown to compare with the simulated current. The two experimental data sets are both recorded for the emitter in Fig. 3.11, but the second experimental curve is taken 11 days (of continuous operation) after the first curve was measured. It is suspected that the curves are slightly different due to changes in the emitter geometry during operation (see Chapter 5).

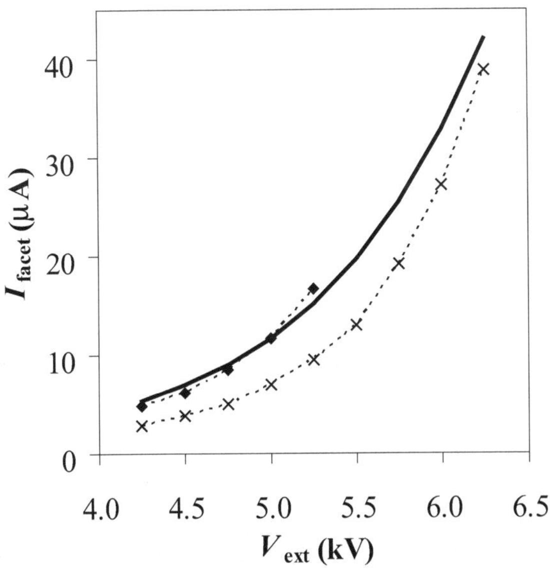

Figure 3.18 Measured (markers) and calculated (solid line) facet currents as a function of the extraction voltage. The measured data are taken at equivalent conditions, but the diamonds show data taken 11 days before the other data series was taken. As will be shown in Chapter 5 the emitter geometry can change during operation.

Figure 3.19 compares measured and calculated angular intensities. The experimental angular intensity data in Fig. 3.19 are taken by the manufacturer in a similar gun geometry for a constant suppressor voltage of −240 V.

It is noted that in general the facet current is a strong function of the field, temperature, work function, and exact shape of the facet and the surrounding emitter area, but it was found that the work function has an almost negligible effect on the exact *shape* of the emission pattern.

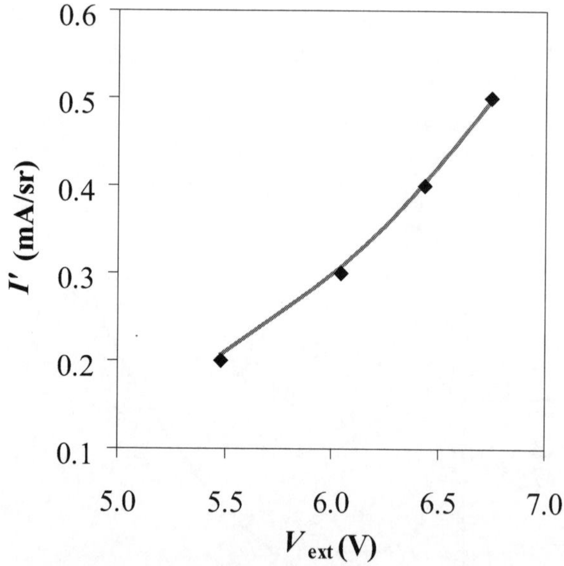

Figure 3.19 Measured (diamonds) and simulated (solid curve) angular intensities as a function of the extraction voltage for a constant suppressor voltage of −240 V. The measured data are from the manufacturer.

As mentioned, for most applications only the emission from the facet *center* is of interest. Also, the angular intensity of the source is often an accessible experimental parameter. Below we discuss properties for the central beamlet in more detail. Figure 3.20 gives the field strength at the facet center as a function of the extraction voltage, and the lens effect for the beamlet from the facet center.

In Fig. 3.20 the field strength at the facet center is a linear function of the extraction voltage, but it does not go to zero at zero extraction voltage due to the presence of the suppressor at the fixed potential. The field enhancement at the facet center, $\beta = F/V_{ext}$, is thus not a constant but depends on the extraction voltage, $\beta = \beta(V_{ext})$. Similarly, the lens effect r/α is not constant, but $r/\alpha = r/\alpha(V_{ext})$. As can be seen, increasing the voltage increases the field strength and the beam divergence. If the field form is kept constant (dashed lines in Fig. 3.20) the change in extraction voltage is directly proportional to the change in the field at the facet. This is exploited in so-called "Schottky plots," which will be discussed in Section 3.4.

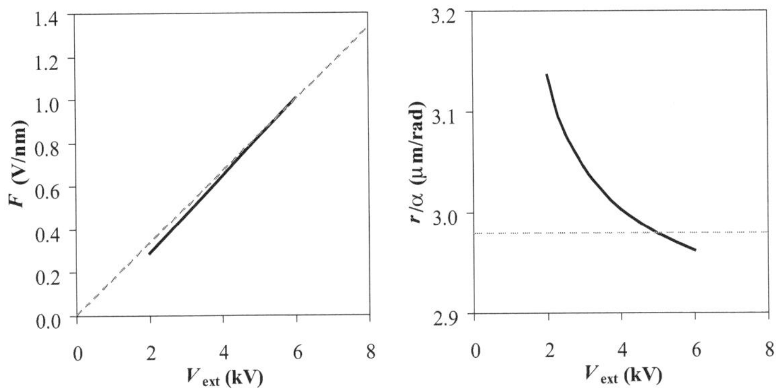

Figure 3.20 Field strength at the facet center (left) and the lens effect experienced by the beamlet from the facet center (right) as a function of the extraction voltage. Dashed lines indicate the field strength and lens effect that would be experienced if the suppressor voltage would be changed with the extraction voltage proportionally (constant field form).

The results in Fig. 3.20 can be compared to the equations in *Handbook of Charged Particle Optics* (Swanson & Schwind, 1997). These Eqs. (3.10) and (3.11) are based on finite-difference simulations (in spherical coordinates with increasing mesh size [SCWIM]) for a sphere-on-cone geometry for which the sphere is truncated by a facet with a diameter of $0.6r_{sphere}$, which determines the angle at the facet to be 17º. For this geometry it was found that the field enhancement at the facet center ($\beta = F_{facet\ center}/V_{ext}$) can be calculated from

$$\beta = 8.33\frac{L_{sup\text{-}ext} - \left(1 - \frac{V_{sup}}{V_{ext}}\right)(L_{tip\text{-}ext} - 6.8 \times 10^{-5})}{L_{sup\text{-}ext} \times r^{0.758}}, \tag{3.10}$$

with $L_{sup\text{-}ext}$ the distance between the suppressor and the extractor, $L_{tip\text{-}ext}$ the distance between the tip and the extractor, V_{sup} the suppressor voltage, V_{ext} the extraction voltage, and r the tip radius.

And it was found that the relation between the lens effect (r/α, or r_{tip}/m_α) and the tip radius and the field enhancement factor can be written as

$$\frac{r_{\text{tip}}}{m_\alpha} = \frac{r_{\text{tip}}^{1-\tau}}{\eta\beta^\tau}, \tag{3.11}$$

with $\eta = 0.525$ and $\tau = 0.42$.

It can already be seen that the tip of the emitter in Fig. 3.11 does not resemble a truncated sphere, but to really compare the equations with the simulation result for the tip geometry in Fig. 3.11 we need to assign a tip radius to the emitter. The emitter radius as specified by the manufacturer is 0.85 μm. This is the larger dashed circle in Fig. 3.21. For this radius the agreement between Eqs. (3.10) and (3.11) and the result in Fig. 3.20 is rather poor. For example, with Eq. (3.10) the field enhancement factor for an extraction voltage of 5 kV is 1.42×10^5 1/m, while Fig. 3.20 gives 1.66×10^5 1/m. To match, the radius should be 0.72 μm.

But the voltage dependence in Fig. 3.20 and that in Eqs. (3.10) and (3.11) is almost equivalent. The agreement between Eq. (3.11) and the right graph in Fig. 3.20 can be made 100% if we use the field enhancement factor from the simulations in Fig. 3.20, with a tip radius in Eq. (3.11) of 0.59 μm. This is the smaller of the circles in Fig. 3.21.

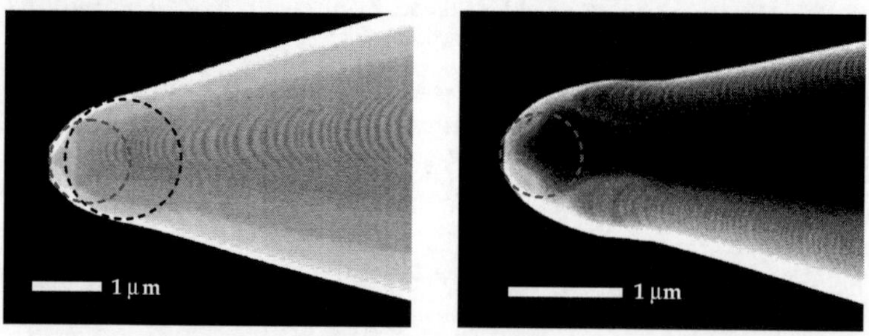

Figure 3.21 SEM images of two different Schottky emitters. Both images contain a circle the radius of which gives the correct lens effect for the central beamlet using Eq. (3.11). The left emitter is as purchased, and it contains a second, larger circle, which is the emitter radius as specified by the manufacturer.

In practice Schottky emitters show a variety of tip geometries. For the other emitter in Fig. 3.21, with a slightly different shape, an agreement between simulations and Eq. (3.11) could be obtained for the radius of the circle that was the best fit of the contour of the emitter near the facet, as indicated—also for this radius no agreement of the simulations with Eq. (3.10).

This shows that for real emitter geometries the definition of the radius to be used in Eqs. (3.10) and (3.11) is ambiguous and that the combination of the two equations does not necessarily apply.

The exact relation between the applied voltage and the field at the emitting surface, and how the emission gets distorted between the emitting surface and the extractor plane (where we can measure it), is a function of the tip geometry. This is further investigated in the next section.

3.3 The Effect of Emitter Geometry

Schottky emitters can have different sizes and geometries, and this affects the lens between the emitter and the extractor. We will investigate the effect of a different tip end, a different tip size, and a different cone shape. The results will be used later in Chapters 4 and 5.

3.3.1 Tip End

The difference in lens effect between a spherical tip end and a truncated spherical tip was shown by Kang *et al.* (1983). They compared a spherical tip end with a *truncated* spherical tip end (∅ facet $0.6r_{tip}$) with the same sphere radius (1 μm), in the same gun geometry, and at the same potentials and found for r/α for the central beamlet a value of 1.8 μm/rad for the spherical tip end, while the tip with the facet yielded a value of 4.2 μm/rad. Figures 3.23–3.25 give an example of the lens effect on the emission properties of a different facet size, as indicated in Fig. 3.22.

The simulations are performed for the standard gun geometry with a distance between the suppressor and the extractor of 0.77 mm, and a protrusion of 292 μm, for a suppressor voltage of −0.3 kV and an extraction voltage of 3.5 kV with respect to the emitter, a temperature of 1,800 K, and a work function of 2.93 eV.

The results show that at constant voltage settings the field strength at the facet center, and the divergence (r/α) of the beam from the facet center, decrease with increasing facet size. As a result the current density on the largest facet is the lowest (lowest field), but the angular current density is highest (less divergent beam).

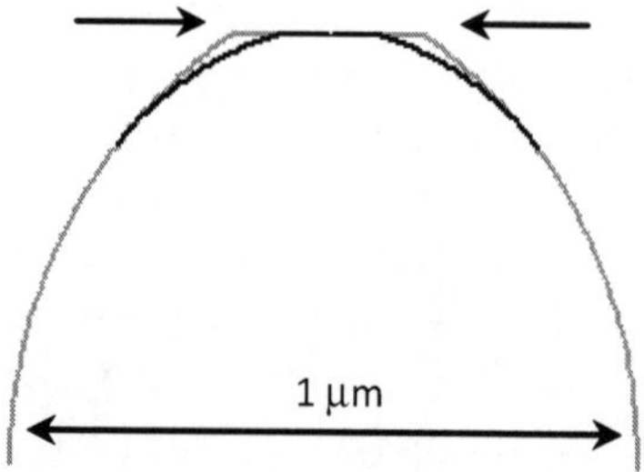

Figure 3.22 Tip geometry with varying facet size. Results in Figs. 3.23–3.25 are based on the given contours and the contours in between.

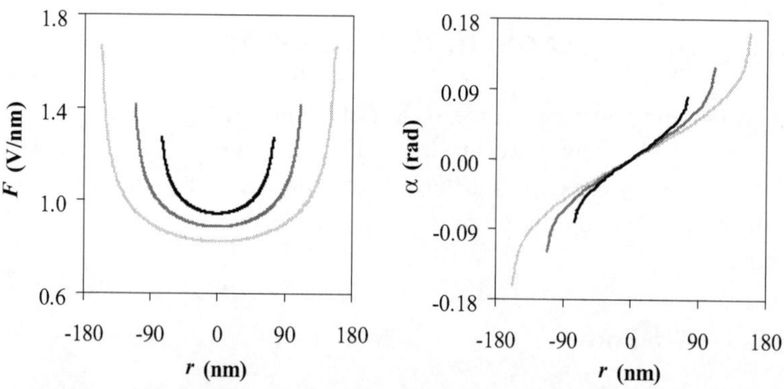

Figure 3.23 Field strength across the facet and lens effect for different launch positions on the facet. The black curve corresponds to the contour in Fig. 3.22 with the smaller facet, the light gray to the other contour, and the dark gray curve to a contour in between.

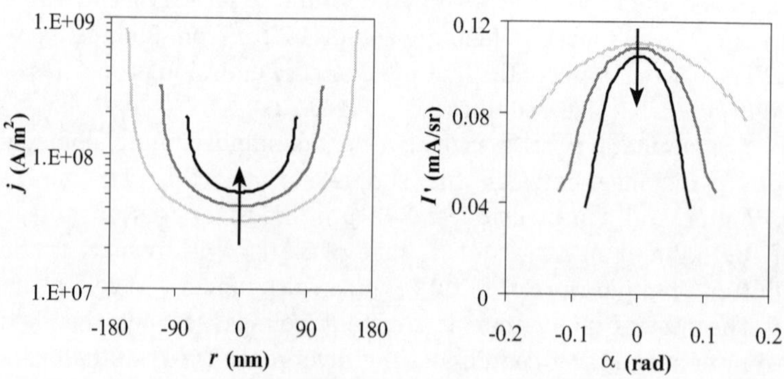

Figure 3.24 Current density across the facet and angular current density behind the extractor.

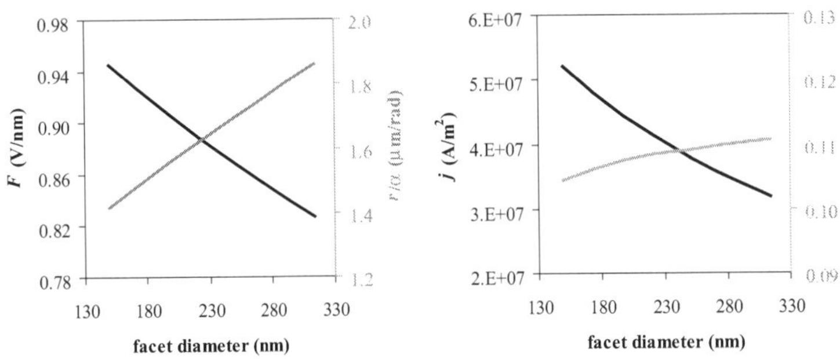

Figure 3.25 Field strength for the facet center and the lens effect for the central beamlet.

While Figs. 3.23–3.25 have illustrated the effect of a different facet size, Fig. 3.26 gives an example of the effect of a different shape of the tip end near the facet. The simulation results correspond to the standard gun geometry with a distance between the suppressor and the extractor of 0.76 mm, a protrusion of 0.26 mm, a suppressor voltage of −0.3 kV and an extraction voltage of 5.5 kV with respect to the emitter, and a temperature and work function of 1,800 K and 2.93 eV, respectively.

Note that while the facet is exactly the same, the size and shape of the emission pattern are quite different.

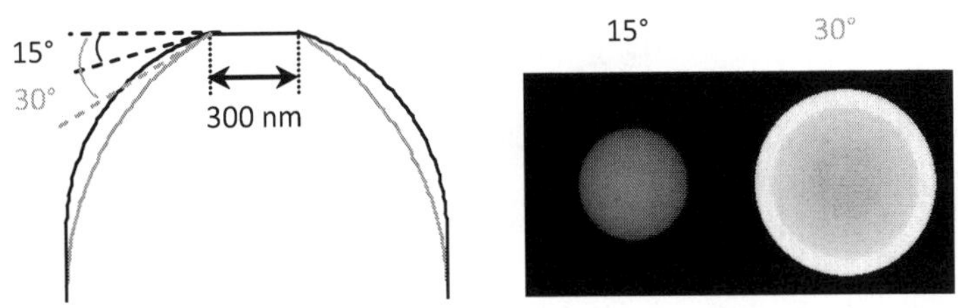

Figure 3.26 Facet emission patterns (same "camera settings") simulated for given tip-end geometries. The field at the facet center is 0.93 (15°) and 1.08 V/nm.

3.3.2 Tip Size

Schottky emitters can be bought in different sizes, and emitter tips can also change in size during their lifetime (Chapter 5). Figures 3.28–3.30 show the effect on the emission properties of a different tip size, as indicated in Fig. 3.27. For the tip end of these emitters only the scale varies, not the geometry.

The simulations are performed for the standard gun geometry with a distance between the suppressor and the extractor of 0.75 mm, a protrusion of 0.25 mm, a fixed suppressor voltage of –0.3 kV with respect to the emitter, an emitter temperature of 1,800 K, and a work function of 2.93 eV. The extraction voltage was varied to establish 1 V/nm at the facet center for all tips.

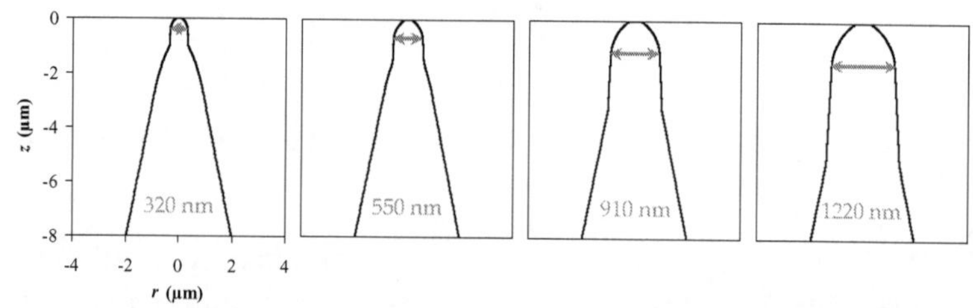

Figure 3.27 Geometries used to investigate the effect of a different tip size on the emission properties of the facet. The tip size is given and corresponds to half the arrow lengths.

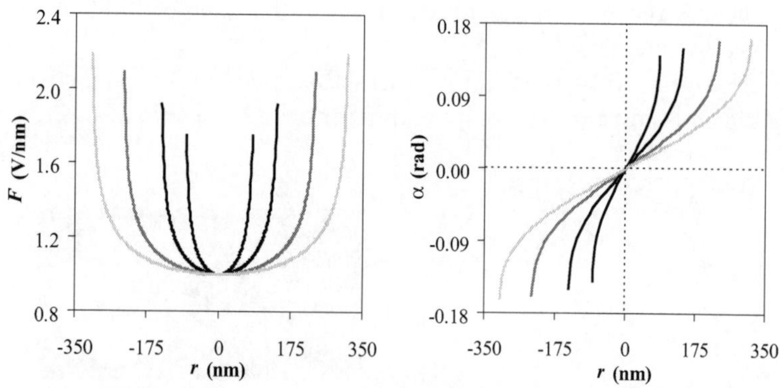

Figure 3.28 Field strength (left) across the facets of the emitters in Fig. 3.27 and the facet extractor lens effect (right). Black curves correspond to the smallest emitter.

Figure 3.30 gives the relation between the tip size and the lens effect for the central beamlet. Note that while the field and the current density at the facet center are equivalent, the angular intensity is 3, 7, and even 11 times higher than that for the smallest emitter in order of increasing tip size.

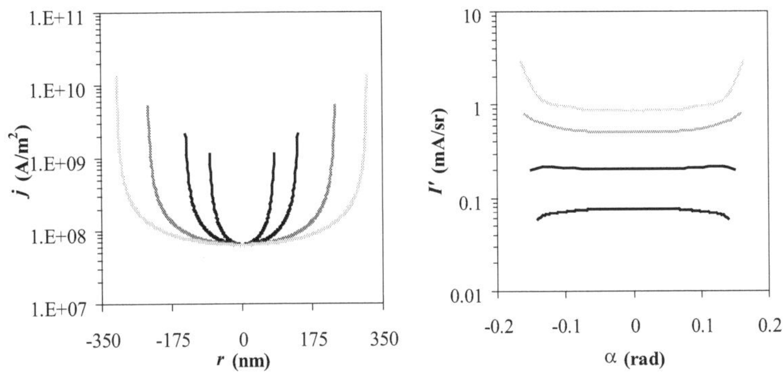

Figure 3.29 Current density across the facets (left) and the angular current density distribution behind the extractor (right). Black curves correspond to the smallest emitter in Fig. 3.27.

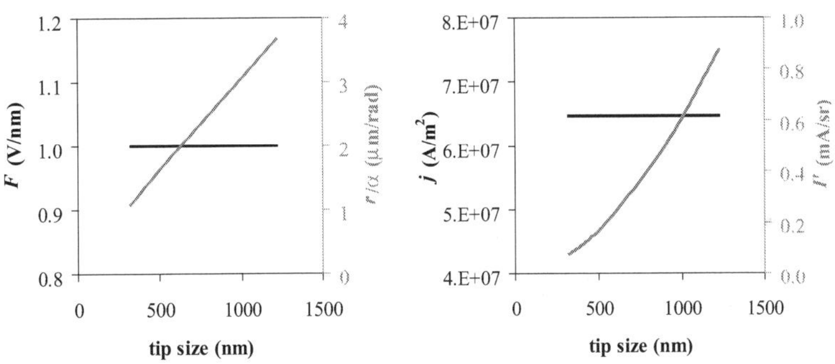

Figure 3.30 The effect of the tip size (see Fig. 3.27) on the emission properties for a constant field of 1 V/nm at the facet center.

3.3.3 Cone Shape

The cone of commercial Schottky electron emitters typically looks like that in Fig. i.1, but other shapes are also possible (e.g., Sakawa *et al.*, 2003). Below we compare the effect of the cone shape on emission properties. The two geometries investigated are shown Fig. 3.31. The results are in Figs. 3.32 and 3.33.

The simulations have been performed for the standard gun geometry with a suppressor–extractor distance of 0.75 mm, a protrusion of 0.25 mm, a suppressor voltage of −0.3 kV and an extraction voltage of 5 kV with respect to the emitter, an emitter temperature of 1,800 K, and a work function of 2.93 eV.

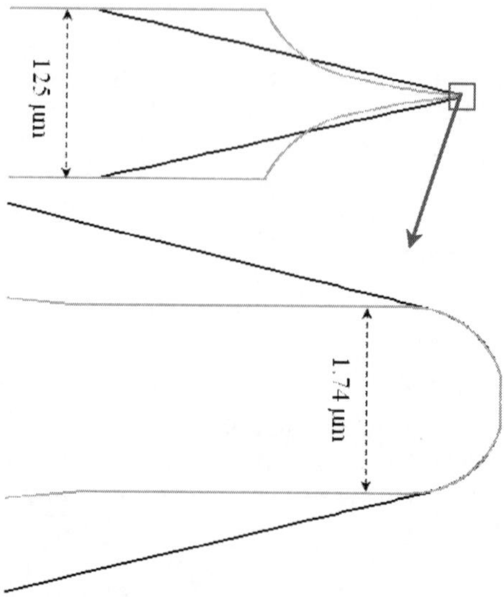

Figure 3.31 Geometries used to investigate the effect of cone geometry on the emission properties of the facet.

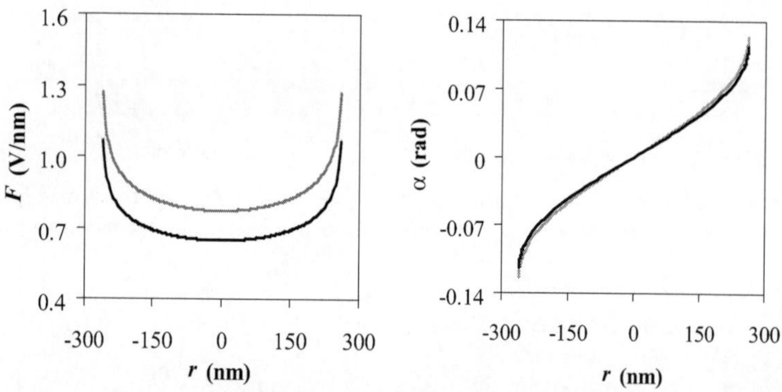

Figure 3.32 Field strength across the facet (left) and facet extractor lens effect (right). Black curves correspond to the black contours in Fig. 3.31.

Figures 3.32 and 3.33 show that for the same voltage settings the emitter with the smaller cone angle near the tip (gray contours in Fig. 3.31) has a higher field and angular current density. The central beamlet of this emitter is, however, more diverging, with an r/α value of 3.4 μm/rad versus 3.7 μm/rad for the emitter with the straight cone (slope at $r = 0$ in the right plot of Fig. 3.32).

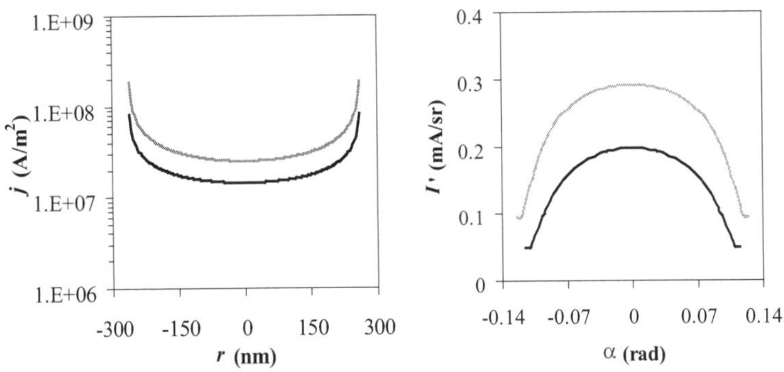

Figure 3.33 Current density across the facet (left) and angular current density behind the extractor (right). Black curves correspond to the black contours in Fig. 3.31.

Due to this difference in the lens effect, for the same *field,* the angular intensity of the emitter with the smaller cone angle is actually 20% lower than the emitter with the larger cone angle (this requires an extraction voltage of 4.3 kV for the "small cone angle" emitter). For the same properties at the facet center, the emitter with the lower cone angle thus has a lower angular current density.

In conclusion, emitters of different geometries that display the same angular intensity may have very different conditions at the emitting surface. Or, emitters with equivalent conditions at the emitting surface may have very different angular intensities. A single equation as Eq. (3.10) to capture all possible variations that are encountered in practice (tip size, facet size, angle at the facet, cone angle) is currently not available, and it would require a lot of simulation work to derive it.

In general, a single measurement of the angular intensity of an emitter for which the tip geometry is not known exactly is thus not very informative. However, a series of measurements of the angular intensity for different extraction voltages, can tell a lot more. This will be discussed next.

3.4 Schottky Plots

In practice often the emitter gun geometry is not known exactly, for example, because the gun geometry is slightly altered upon mounting or after baking, or because the emitter has changed its shape during operation and this has not been monitored. In that case the lens effect is unknown and a measured angular intensity does not give any information on the situation at the facet. But a series of measurements of the angular intensity at different extraction voltages can actually yield the field enhancement factor at the facet center, that is, when the temperature of the source is known. To know the field strength is useful for a shape stability analysis, and the field strength is one of the input parameters for calculation of emission properties (Chapters 1 and 4).

The theoretical ground for the method to find the field enhancement factor, is the analytical equation for Schottky emission, Eq. (1.28), in which tunneling is neglected. Equation (3.12) gives the natural logarithm of the current density:

$$\ln j_S = \underbrace{\ln\frac{em}{2\pi^2\hbar^3}(k_B T)^2 - \frac{\phi}{k_B T}}_{\text{axis intercept}} + \underbrace{\frac{\sqrt{\frac{e^3}{4\pi\varepsilon_0}}}{k_B T}}_{\text{slope}}\sqrt{F} \tag{3.12}$$

From Eq. (3.12) it can be seen that a plot of the natural logarithm of the current density versus the square root of the field strength will be a straight line if the work function and temperature can be assumed to be constant. Experimentally, however, $I'(V)$ is measured and not $j(F)$. But if $I' = f^2 j$, Eq. (3.5), $V = F/\beta$, and f and β are independent of the voltage settings (constant field form), the plot of $\ln(I')$ versus $\sqrt{V}$ will also yield a straight line:

$$\ln I' = \underbrace{\ln\frac{em}{2\pi^2\hbar^3}(k_B T)^2 - \frac{\phi}{k_B T} + 2\ln f}_{\text{axis intercept}} + \underbrace{\frac{\sqrt{\frac{e^3}{4\pi\varepsilon_0}}}{k_B T}}_{\text{slope}}\sqrt{\beta}\sqrt{V} \tag{3.13}$$

As can be seen from Eq. (3.13) the slope of the experimental plot can give the field enhancement factor β (if the operating temperature is known). Such a plot, of $\ln I'$ versus $\sqrt{V}$, is a so-called Schottky plot. The field enhancement factor can be found from

$$\beta = \frac{4\pi\varepsilon_0}{e^3}(k_B T)^2 \text{slope}^2 = 5.157(T \times \text{slope})^2. \tag{3.14}$$

If the emitter geometry is known exactly, the lens effect can be found from simulations, and in that case the experimental Schottky plot can also give the work function

$$\phi = k_{\mathrm{B}}T\left(\ln\frac{em}{2\pi^2\hbar^3}(k_{\mathrm{B}}T)^2 + 2\ln f - \text{axis intercept}\right). \tag{3.15}$$

For an emitter of known geometry (known β and $f = r/\alpha$) and temperature the axis intercept of a Schottky plot can thus be used to check the work function of the surface, for example, if there is reason to believe that the emitting surface might be contaminated or depleted from ZrO_x.

(But also, if the work function is known, Eq. (3.15) could be used to find the lens effect.)

A requirement for a correct Schottky plot is that for each voltage setting the electrons collected by the aperture originate from the same area on the facet and the properties for this area (such as field strength) be uniform across that area. Probing the same area for each voltage setting can be realized with a Faraday cup on the optical axis of a properly aligned gun and a constant field form around the emitter for the different extraction voltages. It is noted that perfect alignment of the emitter with respect to the suppressor and extractor has been proven difficult. For example, FEI Company* specifies for each emitter supplied with a suppressor cap the centering of the emitter with respect to the suppressor aperture, which is usually ~0.01 mm. This is at room temperature. Upon heating the emitter the centering might change due to asymmetry in the construction of the emitter and heating filament.

The constant field form can be achieved by varying the suppressor voltage proportional to the extraction voltage. Because in the past Schottky plot data have often been taken for constant suppressor voltage, the effect on the plot of a constant or a changing field form is illustrated in Fig. 3.34. These are artificial Schottky plots for the emitter in Fig. 3.1, in the standard gun geometry of Fig. 1.2, with a suppressor–extractor distance of 0.75 mm and a protrusion of 242 μm, assuming a temperature of 1,800 K and a work function of 2.93 eV. The difference in the slope of the two curves yields for this example a difference in the field enhancement factor of 16% (which translates to a difference in the calculated current density of 86%).

---

*FEI Beam Technology Division, 5350 NE Dawson Creek Dr., Hillsboro, OR 97124, USA.

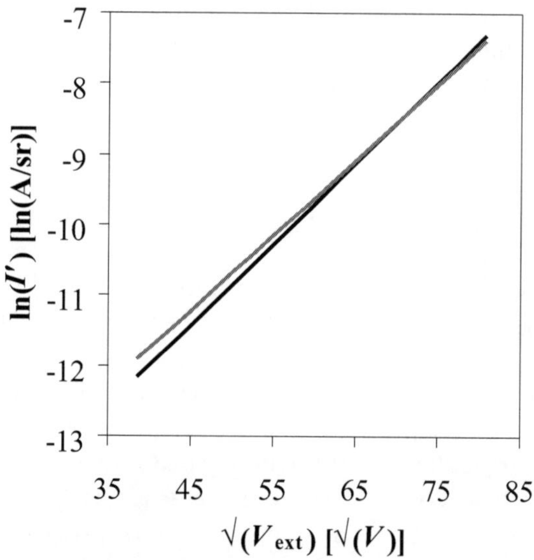

Figure 3.34 Artificial Schottky plot for the emitter in Fig. 3.1, generated with the analytical Schottky emission equation, Eq. (1.28), for a constant field form (gray) and a variable field form (constant suppressor voltage).

A requirement for a correct result is also that Eq. (1.28) be a good approximation. Figure 1.11 has shown that at higher fields the contribution from tunneling becomes significant. This restricts the field range for which Eq. (1.28) is valid. In Fig. 1.11 the tunneling starts to make a difference from ~1 V/nm.

Another aspect is the reflection at the metal–vacuum interface, which is ignored in the analytical model. Figure 3.35 gives a theoretical Schottky plot ($\ln(j)$ vs. $\sqrt{F}$ instead of $\ln(I')$ vs. $\sqrt{V}$), comparing the analytical model with the numerical model. The graph on the right-hand side shows the small deviations of the numerical result with respect to the analytical line.

In practice one cannot measure the whole range of fields displayed in Fig. 3.35. The minimum field is limited by the lowest angular intensity that can be measured accurately and with respect to an upper limit: raising the extraction voltage to a much higher value than the source has been operating at is generally not desirable as the suddenly increased shank emission for higher voltages can induce electron stimulated desorption from the extractor plate, which could cause unstable emission. The minimum field for which a data point for the Schottky plot can be measured depends on the angular intensity of the source in combination with the acceptance angle of the aperture. In general the minimum field will be lower for a larger emitter because for an equivalent field it has a relatively higher angular intensity than a smaller emitter (Fig. 3.30).

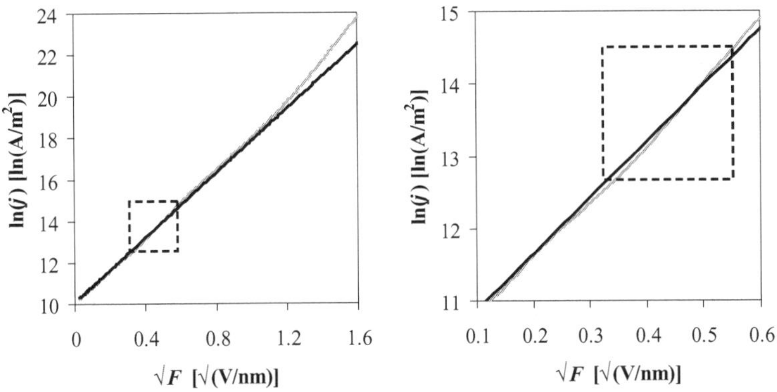

Figure 3.35 Theoretical Schottky plots comparing the numerical calculation (gray) and the result from the analytical Schottky emission equation (black). T = 1,800 K; ϕ = 2.93 eV. The right image is a close-up of (part of) the left image. The window for which Schottky plot data can be measured in practice is limited. The dashed square indicates a typical field range.

From the gray curve in the right image of Fig. 3.35 one can imagine that if the measured field range is limited, the slope of a fit with a straight line, Eq. (3.13), will depend on which part of the curve is measured. A specific example is given in Fig. 3.36. Suppose $I(V)$ data are taken for a voltage range of 1,000 V around the operating voltage with a step size of 100 V. The field form is kept constant, and the emitter has a field enhancement factor of 2×10^5 1/m. The figure shows the deviation in the calculated field enhancement factor and work function for a voltage range shifting along the extraction voltage axis.

The question is to what extent the numerical calculations, based on the potential barrier in Eq. (1.15) and truncated as in Fig. 1.6, can approximate emission from real Schottky emitters. This is investigated with Fig. 3.37. Figure 3.37 gives experimental Schottky plots recorded for three different emitters serving in two different scanning electron microscopes (procedures in appendix A). The right graph in Fig. 3.37 gives the experimentally found deviation from the analytical Schottky emission line and compares that to the theoretically calculated deviation. The figure illustrates that periodic deviations indeed exist for the Schottky emitter surface, and it also illustrates that in practice only part of the oscillation period can be measured. The theoretical and experimental curves are different, however, both in frequency and in phase. This suggests the potential barrier of Eq. (1.15) is not a perfect description, the free-electron model does not apply, and/or the linear fits in the left graph are not true Schottky lines.

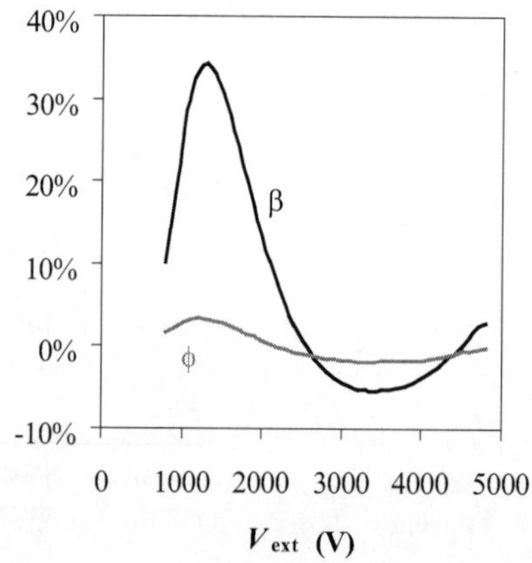

Figure 3.36 Deviation from the true field enhancement factor and work function by fitting artificial Schottky plot data for a constant field form over a 1,000 V range around V_{ext} in steps of 100 V with Eq. (3.12). The input geometry has a field enhancement factor of 2 × 10^5 1/m, and the emission properties are calculated numerically, as described in Chapter 1, with a work function of 2.93 eV and a temperature of 1,800 K.

It is noted that the agreement between the theoretical and experimental curves increases if we assume a much higher experimental temperature (few hundred degrees): this stretches the experimental curves and shifts them to the right. This could suggest the peak of the potential barrier in Fig. 1.15 is too far from the "surface" or that the relation between the electron energy and the wavelength is different than assumed.

Despite all of this, a Schottky plot is still considered a useful tool to find an estimate of the field strength at the center of the end facet on an emitter (provided the data are taken correctly and the operating temperature is known) and to monitor changes of the emitter shape (causing a change in the Schottky plot slope), and it will be used as such in the remainder of this book on many occasions.

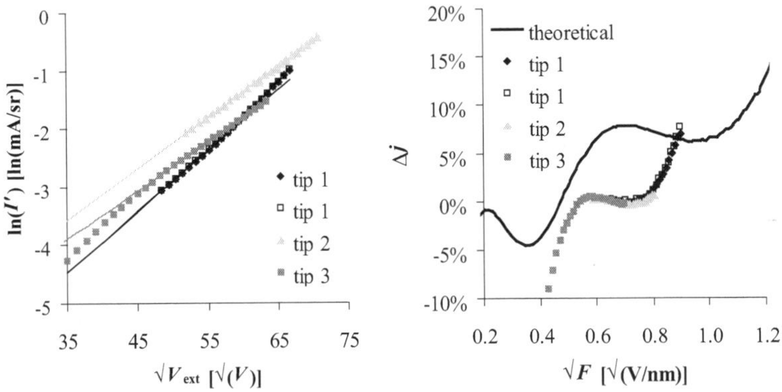

Figure 3.37 Experimental Schottky plots for three different emitters in two different scanning electron microscopes (left) and deviation from the analytical Schottky line (right). The theoretical curve in the right figure is the numerical result for a Fermi energy of 11.47 eV and a temperature of 1,800 K. Assumed temperatures for the experimental curves are 1,800 K for tips 1 and 2 and 1,850 K for tip 3. Experimental data are taken by ramping the extraction voltage down from the operating voltages. Tip 1 is measured twice, with the second measurement a few days after the first. The calculated field enhancement factors are respectively 1.8×10^5, 1.3×10^5, and 1.2×10^5 1/m.

3.5 The Effect of the Heating Current

Increasing the filament current to heat the emitter has many consequences (see also Section 2.3). It changes the Fermi–Dirac distribution for the occupation of the electron energy levels in the material, the (surface) diffusion rates increase, possibly the surface structure changes, the tip shape might change a little bit (Chapter 5), the emitter expands, and the radiation from the emitter increases, which will heat up the emitter environment. All these different effects are expected to change the emission and with different time dependencies.

As an example of the changes in emission upon changing the heating filament, Fig. 3.38 gives the change in the emission pattern and in the facet current ("screen current") upon quickly reducing the heating current, measured for the emitter in Fig. 3.11 in the standard gun geometry, operated with a suppressor voltage of –0.3 kV and an extraction voltage of 6.25 kV with respect to the emitter. Assuming a constant emitter shape and work function, the theory of Chapter 1 predicts a reduction of the emission for all positions on the facet.

In Fig. 3.38, upon reducing the heating current, the intensity in the pattern center starts to decrease, as expected, but the intensity of the pattern edge is initially approximately constant. Also, the reduction in intensity is much less for the pattern edge than for the pattern center. It seems that upon reducing the heating current the facet shape becomes more clearly visible, which is exactly opposite to the effect of reducing the extraction voltage.

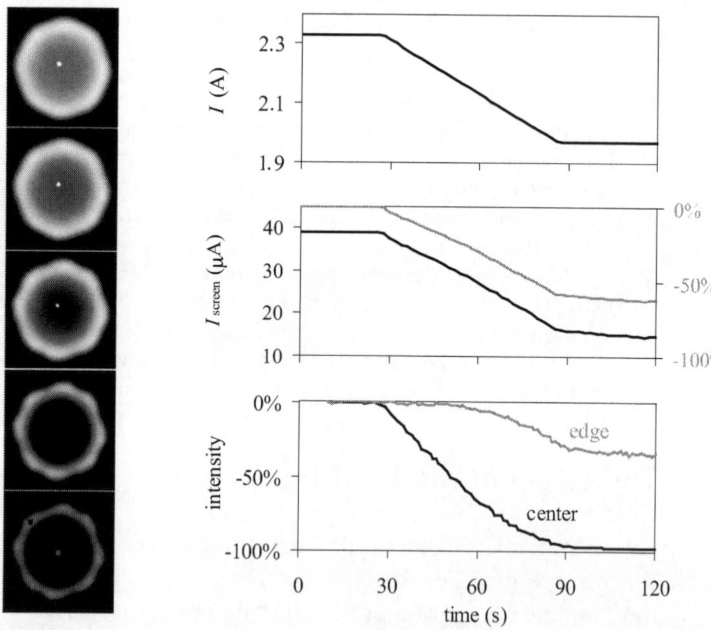

Figure 3.38 Effect of a change in heating current on the emission pattern, the total screen current, and the intensity in the pattern center. The two square-shaped spots in the bottom pattern indicate the position for which the center and edge intensity have been evaluated.

Figure 3.39 gives the response of the emitter on changes in the heating current measured shortly after the data in Fig. 3.38 were taken. This time the emitter is given some time to equilibrate. The changes in heating current are of the order of 50 K (0.04 A).

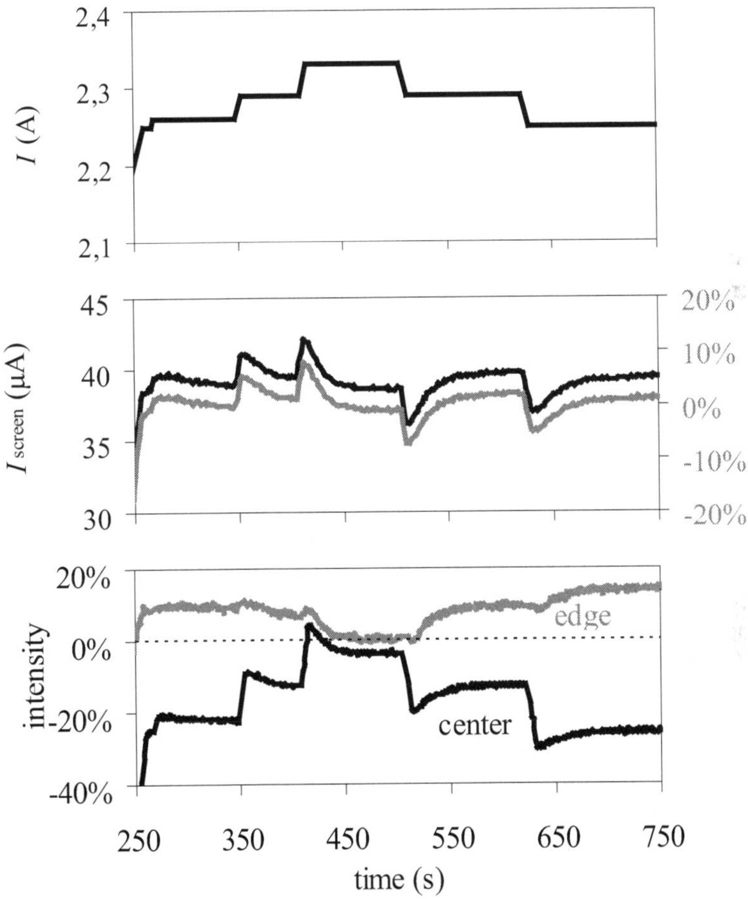

Figure 3.39 Effect of a change in heating current on the facet current and on the intensity in the pattern center and on the edge of the pattern.

First of all it is noticed that the intensity at the pattern center increases with increasing heating current and decreases with decreasing heating current, but the intensity of the pattern edge shows the opposite trend. And as a result, for this tip, in these conditions, the total facet current is approximately independent of temperature.

It can also be seen that upon a change in heating current, the measured current and intensities show some relaxation: for an increase in heating current the current jumps up but then decays, while for a decrease in heating current the current drops but then recovers. Also, the relaxation time seems to increase with decreasing heating current.

The relaxation time is ascribed to a change of the work function of the surface. Changes in tip temperature and protrusion are expected to be much faster, and also, they are expected to give the opposite response. And the change in central intensity is considered too large, and the time scale too fast, to ascribe the behavior to changes of the tip geometry. Figure 3.39 suggests a positive temperature dependence of the work function: it increases with increasing temperature.

It is important to know whether the work function is temperature dependent, because it affects the optimal operating temperature of the emitter: with respect to temperature a trade-off has to be made between emission properties and diffusion and evaporation rates. For higher temperature the emission increases but also the diffusion rates increase, which can accelerate changes of the thermodynamically unstable shape of the emitter (Chapter 5), and the evaporation rate of the ZrO_x reservoir increases, reducing the lifetime of the reservoir that ensures the low work function of the {100} surface. The temperature dependence of the work function is further investigated below.

3.5.1 A Temperature-Dependent Work Function*

In the literature any temperature dependence of the work function of the Zr/O/W{100} Schottky electron source has been largely ignored. Up till recently only Danielson (1981) and Danielson and Swanson (1979) have pointed explicitly at the possibility of a temperature dependence in the relevant temperature regime on the basis of a study of W{100} single-crystalline disks prepared with cycles of Zr evaporation and subsequent oxidation. This yielded a positive coefficient of 0.3 meV/K for a temperature range of 1,450–1,800 K (Danielson, 1981) and 1,000–1,450 K (Danielson & Swanson, 1979) on the basis of extrapolated experimental Schottky plots and a so-called Richardson plot. More recently, for 1,000–1,500 K, Tamura *et al.* (2005) have found a negative coefficient based on the shifts of the onsets of the secondary electron energy distributions. The negative

*Most of this paragraph has been previously published and was reproduced with permission from *J. Vac. Sci. Technol. B*, **24** (2006) 887, Copyright 2006, American Vacuum Society (see also page 249 of this book).

coefficient was associated with a reversible phase transition from a c(2 × 4) + c(4 × 2) double-domain structure at room temperature to a p(1x1) structure at high temperature (Iro96). The results of Kim *et al.* (Kim97a), based on experimental Schottky plots for an emitter at 1330 K, 1,450 K, 1,700 K, and 1,800 K, seem to be consistent with that of Tamura *et al.*, but Kim *et al.* mention that they have let the source equilibrate after every change of extraction voltage until the emission was stabilized. Since the tip shape can change with the extraction voltage whenever the tip shape is not in equilibrium (Chapter 5) this could mean that their Schottky plots contain data for a variable field form and may not give reliable results.

Here we determine the temperature dependence of the work function for temperatures between 1,600 K and 1850 K, operating at ~0.5–0.8 V/nm, using emission patterns and the analytical extended Schottky emission equation, Eq. (1.29).

The emitter that was used was a new emitter from FEI Company with a radius of 0.52 μm (as specified by the manufacturer) and supplied with a suppressor cap and heating wires. The emitter was mounted in the experimental setup described in Section 2.2 (setup A). The background pressure in the system was $\sim 1 \times 10^{-7}$ Pa. The emitter protrusion at room temperature was 0.26 mm. The field at the tip apex was varied by changing the extraction voltage. The suppressor voltage was kept constant at –0.3 kV with respect to the emitter. The temperature of the source was varied by changing the filament current.

The field at the central area of the tip facet was calculated with Eq. (3.10). The temperature-dependent protrusion was neglected. For each temperature and field setting the total current collected on the yttrium-aluminum-garnet (YAG) screen was measured and used to calibrate the intensity scale of the emission pattern in terms of current density. The current density of the central area of the beam at the screen was translated into a current density at the emitting surface using Eq. (3.11).

The work function at each temperature and field was determined by a least squares fit of the experimentally derived current density at the tip surface to the theoretical current density, Eq. (1.29), with the work function as the fitting parameter. For each temperature setting, data were recorded for a series of extraction voltages before the temperature was changed to record the next series. Since the tip shape can change with the applied electric field (Chapter 5) it is important to record the data for different extraction voltages within the typical timescale of any tip shape changes that affect the local field. For the same reason it is important to return to the same field each time before changing the temperature.

The fit results for all settings are given in Fig. 3.40.

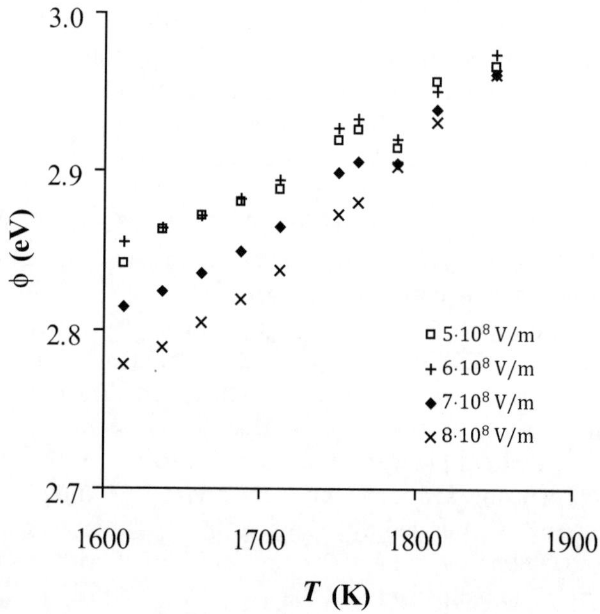

Figure 3.40 Fit results for the work function of a Zr/O/W{100} Schottky electron source at different temperatures and fields.

Figure 3.40 clearly shows a temperature dependence of the work function between 1,600 K and 1,850 K. Possibly there is also a field dependence of the work function: the temperature dependence seems to be stronger for higher fields. Let us first leave out this possibility. A linear fit of the data in Fig. 3.40 then gives (with ϕ in eV)

$$\phi(T) = 6.1 \times 10^{-4} \times (T - 1800) + 2.93\,. \tag{3.16}$$

The statistical errors in Eq. (3.16) are 0.61 ± 0.05 meV/K for the coefficient and 2.93 ± 0.01 eV for the absolute value at 1,800 K.

If the work function is considered to be field dependent too, the best fit for the temperature dependence of the work function for each field is given in Table 3.1. Other results on lower fields yielded temperature coefficients that are within the range given above, but the trend in Table 3.1 is not maintained.

Table 3.1 Fit results for a linear temperature dependence of the work function of a Zr/O/W{100} Schottky electron source operating at different field strengths

F (V/m)	5×10^8	6×10^8	7×10^8	8×10^8
Coefficient (meV/K)	0.51 ± 0.04	0.50 ± 0.04	0.63 ± 0.03	0.79 ± 0.03
$\phi_{T = 1,800}$ (eV)	2.940 ± 0.004	2.944 ± 0.004	2.926 ± 0.003	2.915 ± 0.003

Possible systematic errors are a wrong field enhancement factor, a wrong temperature calibration, and a wrong conversion of the intensity on the YAG screen into a current density on the facet.

The field enhancement factor can be wrong because we have neglected thermal expansion and we have used Eq. (3.10) (derived for a particular tip geometry, and with a tip radius as input parameter, which for real Schottky emitters is ambiguously defined). Because care has been taken to ensure a stable tip shape, a change in the field enhancement factor due to tip shape changes during the experiments is not expected.

A wrong field enhancement factor has two effects. First of all it gives a wrong field value entering the expression for the theoretical current density, Eq. (1.29), that is fitted to the experimental current density to obtain the work function. But it also affects our translation of the current density on the screen to the current density at the facet. A wrong field enhancement factor affects both the absolute work function value and its temperature coefficient, but the possible error in the field enhancement factor is not large enough to wipe out the temperature dependence that was found.

A temperature offset would have a considerable effect only on the absolute work function value but not on the temperature coefficient. The latter, however, does depend on the relation between tip temperature and filament current. This relation has been determined by the manufacturer in a very similar setup. An error of 10% in the temperature range investigated would yield an error of ±0.1 meV/K for the temperature coefficient of the work function.

Finally the conversion of the measured intensity on the YAG screen into a current density on the facet may introduce errors both in the absolute value and in the temperature dependence of the work function. For example, the calculated current density may be too high or too low due to a wrong background correction of the emission pattern, or there is an error in the translation from screen current density to facet current density—based on Eqs. (3.10) and (3.11).

A different temperature dependence is found when the results for the three lowest temperatures (lowest intensities) or the three highest temperatures (highest intensities) are left out of the fitting procedure: while the absolute work function at 1,800 K doesn't change significantly, the temperature coefficient changes to 0.66 meV/K and 0.60 meV/K, respectively. Including a field dependence of the work function, the effect of a wrong conversion is strongest for the highest field in this study, for which the total intensity variation with temperature and the variation of the intensity distribution with temperature is strongest. Leaving out the three lowest temperatures gives a coefficient of 0.89 meV/K and an absolute value at 1,800 K of 2.917, while leaving out the three highest temperatures yields 0.71 meV/K and 2.904 eV. These variations might also have to do with the temperature- and field-dependent reflection of the electrons near the surface, which is neglected in the analytical extended Schottky emission equation.

Not much is known about what exactly will happen to the Zr/O/W{100} surface when the temperature is varied in the range of 1,600 K to 1,850 K. From the literature there is no indication of a change of coverage with temperature or atomic rearrangement on the Zr/O/W{100} surface for this regime. We suggest a few explanations of the measured temperature dependence of the work function that might be plausible. First of all, as many adsorbates, the ZrO_x coverage might be temperature dependent. In theory this can work both ways: the work function might decrease, because there are more dipoles to lower the effective work function, but it might also increase, due to depolarization effects between ZrO_x that reduce the effectiveness of the ZrO. Another explanation is that the Zr/O dipoles might "sink" deeper in the tungsten upon thermal expansion of the bulk. As a consequence the effective work function reduction becomes less effective for higher temperatures, yielding a positive temperature dependence of the work function of the surface. A third possibility lies in the thermal vibrations of the dipoles at the surface, which might cause a reduction of the effective Zr/O dipole moment in the direction normal to the surface (Kuz'min & Mittsev, 2001). Considering the time scale of the response in Fig. 3.39 a change in coverage seems the most plausible.

A possible explanation for a field dependence might be that the dipole moment of the Zr/O dipole is changing with the field applied (Gomer & Swanson, 1963). Because of the limited number of fields investigated and neglect of the temperature-dependent protrusion and the electron reflection at the surface, we do not wish to draw very definite conclusions about a field dependence.

It is noted that the temperature dependence of the work function of the emitting surface of the Zr/O/W{100} electron source we find compares to values for other metal oxides on tungsten (Haas, 1972) but is opposite to and much stronger than that of the bare W{100} surface (Hutson, 1955).

In comparison to Danielson (1981), who has given results for a similar temperature range, we have found a higher coefficient. This might be ascribed to the different circumstances in which the measurements have been performed. While we have investigated a commercial Zr/O/W{100} electron source in standard operating conditions, Danielson has investigated a single-crystal disk and used a different surface preparation. Furthermore Danielson has derived his results from measurements in a lower field range. If there is any field dependence of the dipole moment of the Zr/O dipole, this might explain the different results. A third possibility could be the temperature range investigated by Danielson. Tamura *et al.* have found that below ~1,500 K the work function starts increasing again with decreasing temperature. Since temperature is a difficult-to-measure parameter and Danielson has recorded data in the 1,450–1,800 K range it could be that he included data points for a surface in transition with a higher work function, suppressing the coefficient when fitting the results for the indicated temperature range.

In the remainder of this work we will adopt a temperature coefficient of 0.6 meV/K.

We can now explain the response of the emission pattern to a change in temperature, as seen, for example, in Fig. 3.38.

3.5.2 The Predicted Effect on the Emission Pattern

Using the result from above, the effect of the heating current on the emission pattern can now be predicted and compared with experimental patterns. First of all, Fig. 3.41 gives the theoretical current density on the emitter surface as a function of field strength for different temperatures, including the temperature coefficient of the work function of 0.6 meV/K, assuming a work function of 2.93 eV at 1,800 K. Note that as a result of the temperature dependence the curves are not merely shifted but display a crossover, in this case at a field strength of around 1.8 V/nm: for lower fields the current density increases with increasing temperature, while for higher fields the current density actually decreases with increasing temperature.

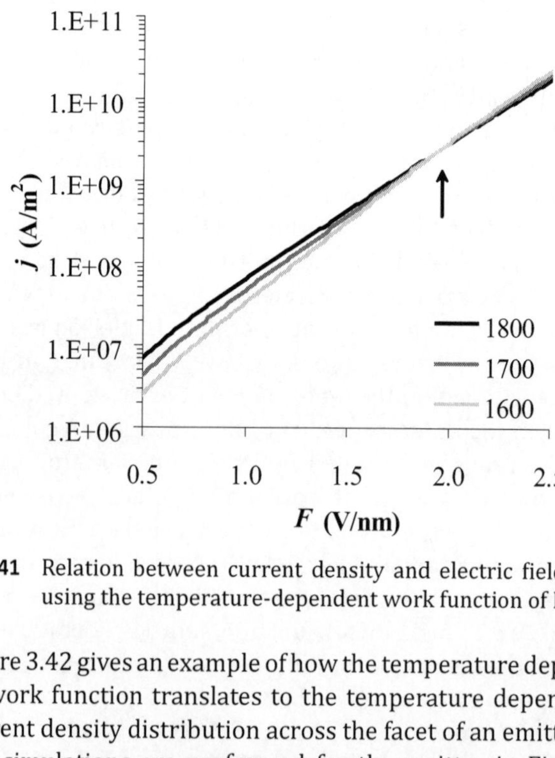

Figure 3.41 Relation between current density and electric field strength using the temperature-dependent work function of Eq. (3.16).

Figure 3.42 gives an example of how the temperature dependence of the work function translates to the temperature dependence of the current density distribution across the facet of an emitter.

The simulations are performed for the emitter in Fig. 3.11 in the standard gun geometry (Fig. 1.2) with a suppressor–extractor distance of 0.76 mm, a constant protrusion of 0.28 mm, and a constant suppressor voltage of −0.3 kV with respect to the emitter. The extraction voltages are 4.25 kV, 5.00 kV, and 6.00 kV, and the temperatures are 1,600 K, 1,650 K, 1,700 K, 1,750 K, and 1,800 K. For the work function we use the temperature coefficient of 0.6 meV/K, and at 1,800 K we use 3.04 eV. The change in emitter protrusion for the given temperature range is estimated to be <10 μm, and its effect is neglected in the simulations.

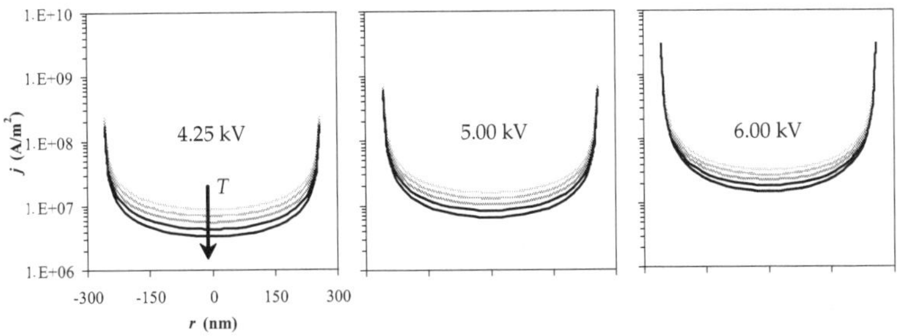

Figure 3.42 Calculated current density across a facet for different temperatures (1,800–1,600 K with a 50 K step, with a work function of 3.04 eV at 1,800 K).

For all applied extraction voltages, the central current density decreases with decreasing temperature. But near the edge of the facet, for the case of 6.00 kV, the local field strength is so high (as can be anticipated from the field distributions across this facet, which were given in Fig. 3.13) that locally the current density slightly increases for decreasing temperature.

Figure 3.43 shows how this translates to angular intensity distributions behind the extractor and how it compares to the experimental patterns. And Fig. 3.44 compares the calculated and measured total facet currents.

The experimental patterns are taken for the emitter in Fig. 3.11 in the standard gun geometry with a suppressor–extractor distance of 0.76 mm, a room temperature protrusion of 249 μm, and a fixed suppressor voltage of –0.3 kV with respect to the emitter.

The agreement between the simulated and experimental emission patterns, and the simulated and measured total facet currents, is considered reasonable. Note that the agreement is poor if the temperature dependence is neglected.

To conclude this chapter, it has been shown that the experimental facet emission pattern and its appearance for different operating conditions is understood and can be reproduced qualitatively and quantitatively.

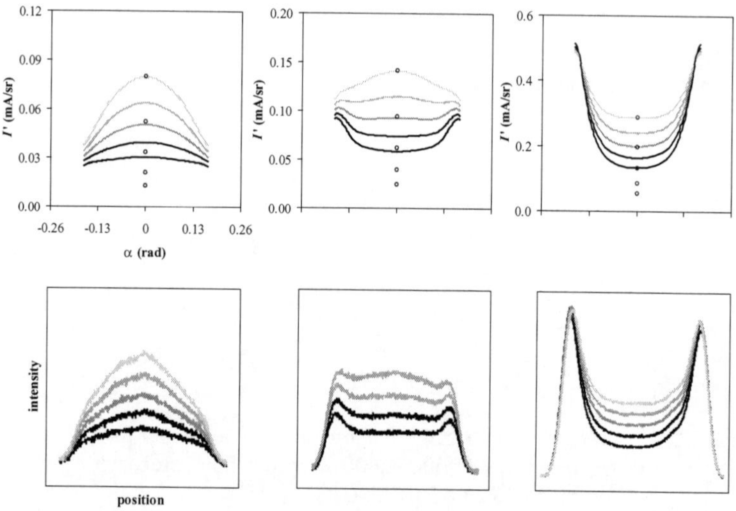

Figure 3.43 Calculated angular intensity distributions (top row) and cross sections of experimental emission patterns at extraction voltages of 4.25 kV, 5.00 kV, and 6.00 kV (left to right) at temperatures of 1,800 K (light gray) 1,600 K (black) (5 kV experimental: 1,750–1,600 K). The markers in the simulated patterns indicate the angular intensity for a temperature-independent work function.

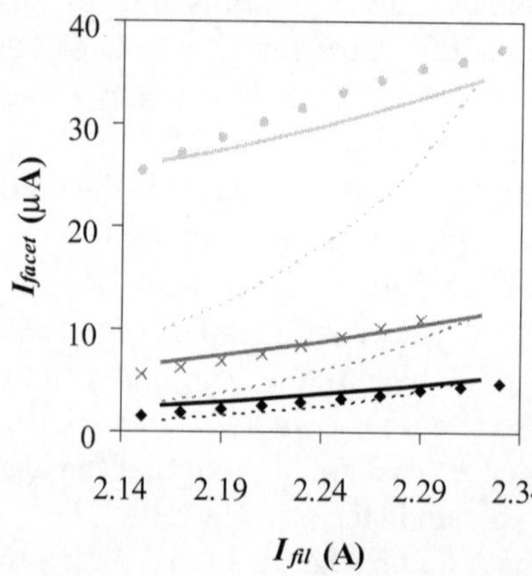

Figure 3.44 Experimental (markers) and calculated facet current as a function of heating current for three different extraction voltages, corresponding to Fig. 3.43. Solid lines: calculated with a temperature-dependent work function; dashed lines: calculated with a temperature-independent work function.

The facet emission pattern can be seen as a distorted "shadow image" of the facet—a shadow because the emission appears to be coming from a point inside the emitter behind the emitting surface (Fig. 3.4) and distorted because the lens between the emitting surface and the extractor has aberrations (e.g., Fig. 3.3). The best "image" of the facet is obtained at a relatively high extraction voltage and low heating current. This will be exploited in Chapter 5.

At the facet the current density distribution across the facet is always very nonuniform, with increased intensity near the facet edge, where the local field is higher, but the current density distribution behind the extractor can be tuned from a convex or concave bell-shaped distribution to a very flat profile by playing with the operating conditions as a consequence of the lens effect. This lens effect is strongly dependent on the emitter geometry and can cause the emission pattern of two different emitters with the same conditions at the facet center to look completely different and give the sources very different angular intensities.

A single measurement of the angular intensity for a Schottky emitter of unknown geometry is thus not very informative about what is going on at the emitter surface. However, from a series of measurements of the angular intensity for different extraction voltages, a Schottky plot, the field enhancement factor can be estimated. The field strength is useful knowledge both for calculating the emission properties of the surface and also with respect to the emitter shape (stability), as will be shown in Chapter 5. First, in the next chapter we will look at the properties of facet emission that are relevant to electron optical systems that use the emission for irradiation of samples. Generally only a very small portion of the total facet emission is actually used for applications. One of the main reasons for cutting the current is that it is not possible to focus the total facet emission to a small spot.

Chapter 4

The Final Beam for Applications

The Schottky emitter is used in systems that make either a probe or a more or less collimated beam to illuminate samples for different purposes. Only a small part of the facet emission is used for this. The column of systems consists of several lenses that make a few intermediate images of the "virtual" source before making the final (de)magnified image onto the specimen (in the case of a probe system) or the final collimated beam (for parallel illumination).

The virtual source is not the physical emitting surface but the virtual crossover formed by the field near the emitting tip. The position and size of the virtual source are of importance to the design of the lens system: they determine the parameters for the optics in between the virtual source and the specimen, such as the lens strength and aperture sizes. The virtual source is addressed in Section 4.1.

A good source for probe formation allows one to squeeze a large current in a small focus spot, while for transmission electron microscopy (TEM) purposes it is associated with a large current in a fully coherent beam. The parameter that determines this ability is the practical brightness. Practical brightness is introduced in Section 4.2.

Diffraction and aberrations make the spot of a focused beam larger than the source image and decrease the spatial (spherical aberration) and temporal (chromatic aberration) coherence of a parallel beam. These different contributions have different

Physics of Schottky Electron Sources: Theory and Optimum Operation
Merijn Bronsgeest

ISBN 978-981-4364-79-9 (Hardcover), 978-981-4364-80-5 (eBook)
www.panstanford.com

dependencies on the opening angle of the beam, which leads to an optimum angle for which the spot size is smallest for a given probe current. The optimum probe current–probe size relation for a given set of source properties will be addressed in Section 4.3.

The chromatic aberration contribution depends on another relevant source parameter: the spread in energies of the electrons in the beam. The practical brightness and energy spread of the beam can be affected by interactions between electrons in the beam. This is the topic of Section 4.4.

Finally, in Section 4.5 the findings of the preceding paragraphs are used to quantify beam properties that are relevant to applications, as a function of the conditions at the emitting surface.

4.1 Imaged by the Electron-Optical System: The Virtual Source

The size and position of the virtual source depend on the source geometry, extraction voltage, temperature, and on which part of the total emitted current is selected for the final probe. The latter depends on the application; for example, for high-resolution imaging the beam current is in the order of picoamperes, but for microanalysis the probe current needs to be much higher. For multibeam applications with off-axis beamlets an additional parameter is introduced: the amount of current is relevant to not only the size and position of the virtual source but also the position of the selecting aperture.

For simplicity we will place the beam-defining aperture (the aperture that selects the probe electrons) right behind the extractor in the field-free zone. In most commercial systems currently available the beam-defining aperture is further away.

The position of the virtual source can be found by tracing the probe electrons back from the field-free zone to the point where they cross the optical axis. Figure 3.4 has already shown that not all trajectories necessarily cross the optical axis in the same point. The position of the plane perpendicular to the axis that contains the smallest full width containing 50% (FW50) diameter of the extrapolated trajectories of the probe electrons is the position of the virtual source.

Most electrons emitted by an operating Schottky emitter have zero tangential energy. Let us first consider an imaginary cold field emitter, which only emits electrons with zero tangential energy, thus perpendicular to the emitting surface. Later we will add the effect of the thermal energy of the electrons. We will investigate a single-emitter geometry that, in Fig. 3.1, is the standard gun geometry with a suppressor–extractor distance of 0.75 mm and a protrusion of 242 μm, a suppressor voltage of −0.3 kV, and an extraction voltage of 5.5 kV with respect to the emitter, unless stated otherwise.

4.1.1 Imaginary Cold Schottky Source

Basically, cold electrons determine the position of the virtual source. When the beam-defining aperture has a small acceptance angle, it can be assumed that the electrons originate from a small surface patch for which the field can be assumed to be constant and for which the lens between the emission site and the extractor is aberration free. As a consequence, the electrons appear to be originating from a single point, which for an on-axis aperture is on the axis (traced back from the aperture) and off the axis for an off-axis aperture, as illustrated in Fig. 4.1. The crossover is the position of the virtual source, and its size is infinitely small.

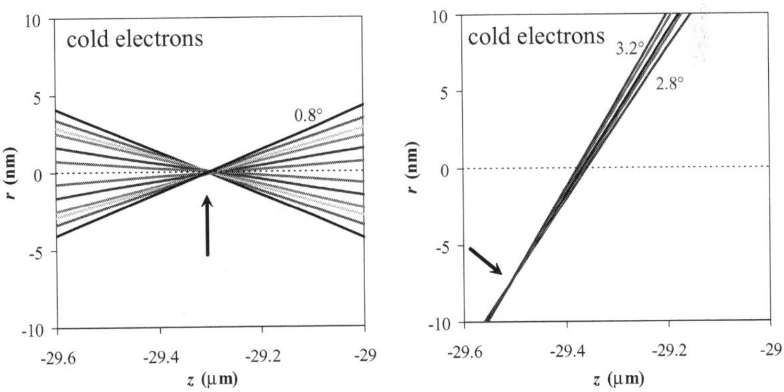

Figure 4.1 Position of the virtual source as found from the trajectories of cold electrons launched from different positions on the facet of the Schottky emitter in Fig. 3.1 and traced back from the field-free zone behind the extractor toward the emitter. $z = 0$ is the position of the facet. The acceptance angle of the beam-defining aperture is 0.8° for the left image and 0.4° for the image on the right. The aperture for the right image is at an angle of 3° with respect to the optical axis.

Figure 4.2 illustrates how the position of the virtual source changes with the position of the beam-defining aperture. Figure 4.3 shows the effect on the position of the virtual source, of a change in extraction voltage at constant suppressor voltage, for a beamlet selected by an on-axis aperture.

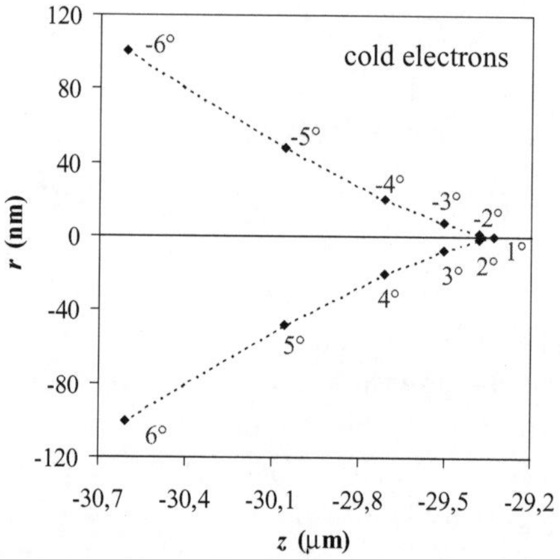

Figure 4.2 Position of the virtual source for a small beam-limiting aperture positioned in the field-free zone behind the extractor, at different angles with the optical axis. $z = 0$ is the position of the facet.

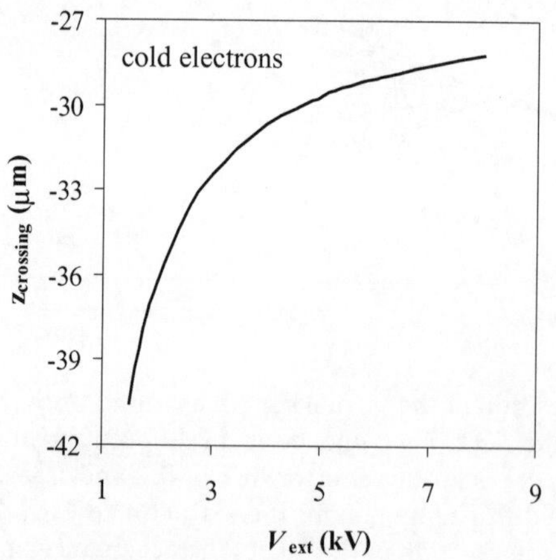

Figure 4.3 Position of the virtual source as a function of the extraction voltage, calculated with a cold electron launched from the facet at 25 nm from the center of the facet of the emitter in Fig. 3.1 (final ray angle 0.7°). $z = 0$ is the position of the facet.

Figure 4.2 shows that the position of the virtual source moves backward and away from the optical axis for increasing distance of the aperture from the optical axis, and Fig. 4.3 shows that the position of the virtual source moves backward with decreasing extraction voltage.

For small beam-defining apertures the size of the virtual source is infinitely small, but for larger acceptance angles of the aperture the virtual source size becomes finite. This is because for larger acceptance angles electrons from a larger emitting area are accepted. For a larger area the field at the emitting surface is no longer the same for all emission sites and also the lens is no longer perfect: tracing back the electrons that reach the aperture it can be seen that they will no longer cross the axis in exactly the same point. This is illustrated in Fig. 4.4.

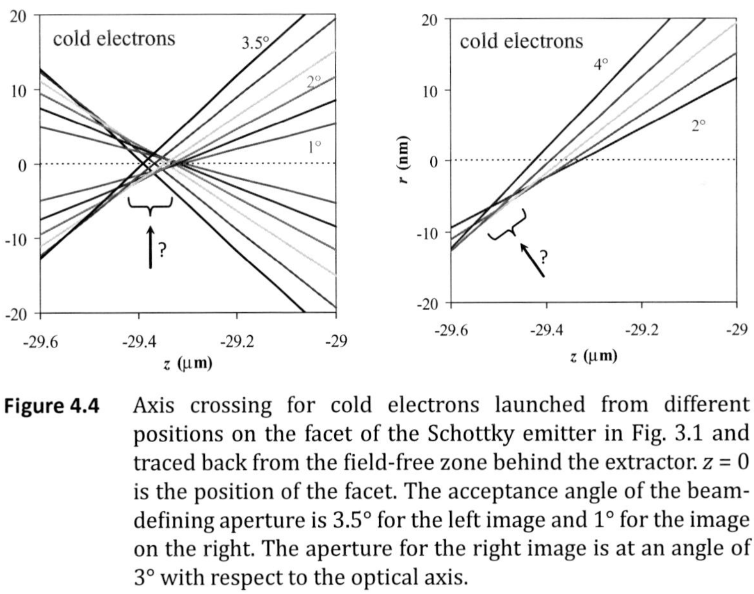

Figure 4.4 Axis crossing for cold electrons launched from different positions on the facet of the Schottky emitter in Fig. 3.1 and traced back from the field-free zone behind the extractor. $z = 0$ is the position of the facet. The acceptance angle of the beam-defining aperture is 3.5° for the left image and 1° for the image on the right. The aperture for the right image is at an angle of 3° with respect to the optical axis.

The virtual source size in Fig. 4.4 is finite due to the aberration. The shape and width of the intensity distribution in the virtual source plane (which determines the FW50) are affected by the aberration and current density distribution on the part of the emitting surface that delivers the beam electrons.

The position of the virtual source is the plane in which the FW50 of the crossover of the beam is minimal.

It is noted that for practical applications generally the acceptance angle of the beam-defining aperture is small enough to ignore the aberration and the virtual source position is determined by the trajectories of the cold electrons.

4.1.2 Heated Schottky Source

In reality the emitted electrons do not all have zero tangential energy but display a distribution of tangential energies. This has been shown in Fig. 1.16. This will give a finite virtual source size, also when the cold electrons make an infinitely small crossover in the virtual source plane.

To find the size of the virtual source for a heated Schottky emitter we will first look at the emission from individual points on the facet and determine their virtual source size. Later we will return to the beam-defining aperture to see which of the electrons emitted by the different emission sites on the facet are accepted by the aperture and how those trace back to the virtual source plane.

Again we will investigate the emitter geometry in Fig. 3.1, in the standard gun geometry with a suppressor–extractor distance of 0.75 mm, a protrusion of 242 μm, a suppressor voltage of −0.3 kV, and an extraction voltage of 5.5 kV with respect to the emitter, unless stated otherwise.

An electron with nonzero tangential energy does not cross the optical axis in the plane of the virtual source, as illustrated in Fig. 4.5. To find the virtual source size we need the relation between the tangential energy of the emitted electron and the point where it crosses the virtual source plane.

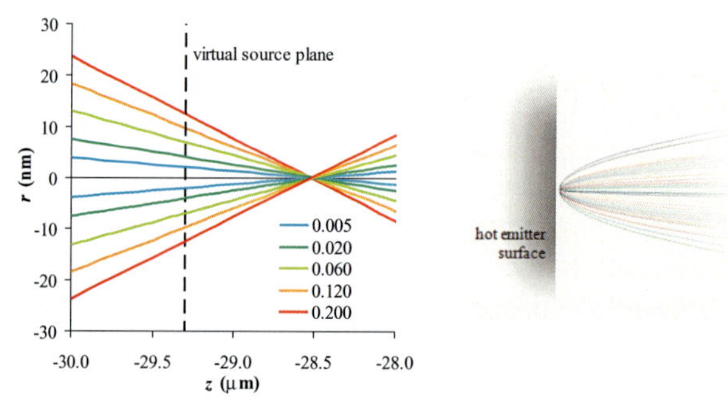

Figure 4.5 Tangents to electron trajectories in the field-free zone behind the extractor, of electrons launched from the facet center with different tangential energies (in eV), make a crossover somewhat before the virtual source plane.

In paraxial approximation, using the conservation of phase space, the distance from the optical axis in the virtual source plane, ρ, of an electron emitted from the facet center with tangential energy E_{tan} can be found from (Law, 1937; Worster, 1969; Wiesner & Everhart, 1973)

$$\underset{\text{emitting surface}}{\left[\Delta r \Delta p_{\text{r}}\right]} = \underset{\text{virtual source}}{\left[\Delta r \Delta p_{\text{r}}\right]} \Leftrightarrow \underset{\text{emitting surface}}{\left[r_{\text{launch}} m v_{\text{t}}\right]} = \underset{\text{virtual source}}{\left[\rho m\left(\alpha_{\text{final}} v_{\text{ext}}\right)\right]} \quad , \quad (4.1)$$
$$\Leftrightarrow r_{\text{launch}} \sqrt{E_{\text{tan}}} = \rho \alpha \sqrt{eV_{\text{ext}}}$$

with v_{t} the tangential velocity and v_{ext} the velocity in the field-free zone behind the extractor. This can be rewritten to give

$$\rho = \frac{r}{\alpha} \sqrt{\frac{E_{\text{tan}}}{eV_{\text{ext}}}} \ . \quad (4.2)$$

Doing this for all tangential energies gives a distribution in the virtual source plane that is basically a projection of the tangential velocity distribution of the emitted electrons. Now we approximate the tangential velocity distribution of the emitted electrons with the Maxwell Boltzmann distribution (see also Section 1.4.3) so that we have a Gaussian distribution for the tangential velocities with a standard deviation of $\sqrt{(k_{\text{B}}T/m)}$. In that case, from Eq. (4.2), the spatial current density distribution in the virtual source plane is also Gaussian and so is the angular current density distribution in the field-free zone behind the extractor. An example is given in Fig. 4.6.

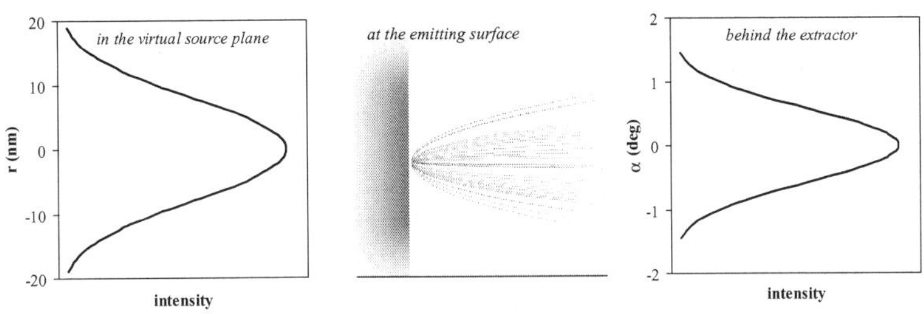

Figure 4.6 Spatial distribution in the virtual source plane (left) and angular distribution in the field-free zone behind the extractor (right) for emission from the facet center of a hot (1,800 K) Schottky source (center).

But in a beam for applications the electrons do not all originate from the facet center. Figure 4.7 shows the projection as in Fig. 4.5 for the center and two other emission sites. It can be seen that the crossover of the tangents to the trajectories of the electrons emitted by different points is at a slightly different position. However, the tangents associated with electrons with equivalent tangential energy all cross the virtual source plane at the same position. This is because the emission sites are not far from the facet center and experience the same lens effect. For emission sites further out from the facet center this will no longer be the case and the effect of aberrations will kick in. For these emission sites also the tangential energy distribution will be different because the field strength for the different sites is not equivalent (e.g., Fig. 3.2).

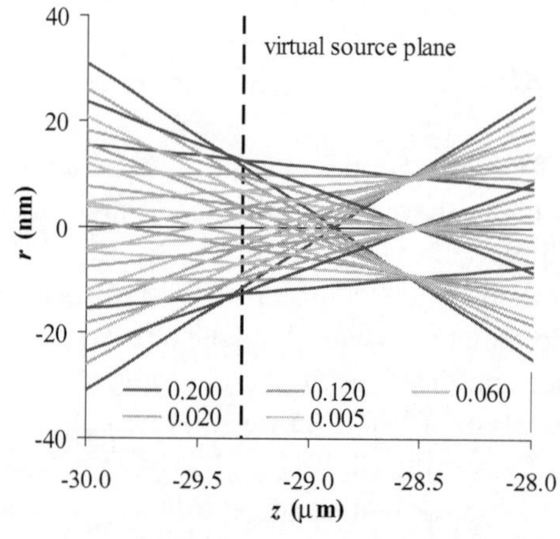

Figure 4.7 Tangents to the electron trajectories in the field-free zone behind the extractor, of electrons launched from the facet center, and from positions 25 nm from the facet center, with different tangential energies (in eV), back-extrapolated to the virtual source plane.

Figure 4.8 compares the tangential energy for spots on the facet 50 nm and 100 nm from the center and also gives the associated angular intensity distribution in the field-free zone.

For sites at 50 nm and 100 nm from the facet center there is more emission, and due to the different lens effect the angular distribution emitted by a spot at 50 nm or 100 nm from the facet center is slightly wider. Note that by adding up the angular distributions of all emission sites on the facet we would get the typical facet emission pattern discussed in Chapter 3. The difference is that in Chapter 3 only the electrons emitted with zero tangential energy were taken into account.

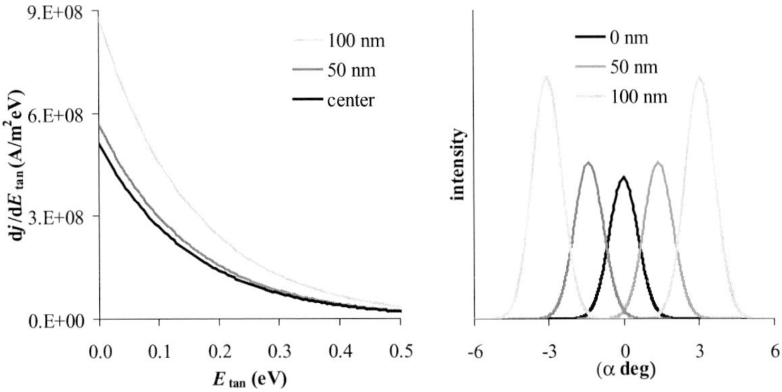

Figure 4.8 Tangential energy distributions of electrons emitted from different positions on the facet (in nm from center) of the Schottky emitter in Fig. 3.1 and the associated angular intensity distributions in the field-free zone. $T = 1{,}800$ K; $\phi = 2.93$ eV.

So far we have been looking at emission from individual sites. To determine the virtual source size for the probe in a system we need to know which electrons from which emission sites are accepted by the beam-defining aperture and to what extent the electrons from emission sites with different properties than the facet center actually contribute.

We put a small aperture at different positions in the field-free zone and see which of the emitted electrons are selected by this aperture. The result is given in Fig. 4.9. It gives the angular intensity distribution accepted on a point on the axis at 2.5 mm from the emitting surface, at a point at 1.4° from the axis, and a point 3.1° from the axis.

It turns out that for the emission collected by the on-axis point, the contribution from emission from sites outside the central facet area is very small. Therefore also for beams for applications the spatial current density distribution in the virtual source plane can be written as a Gaussian

$$J_{\mathrm{v}} \propto e^{-\left(\frac{\rho}{\rho_{\mathrm{T}}}\right)^2}, \tag{4.3}$$

where ρ_{T} is the distance from the optical axis in the virtual source plane of an electron with the mean tangential energy $\langle E_{\mathrm{tan}} \rangle$. We define the virtual source size as the diameter d_{v} of the disc that contains 50% of the total current in J_{v} (FW50):

$$d_v = 2\rho_{FC=0.5} = 2\frac{r}{\alpha}\sqrt{\frac{E_{\tan}^{(FC=0.5)}}{eV_{\text{ext}}}}. \tag{4.4}$$

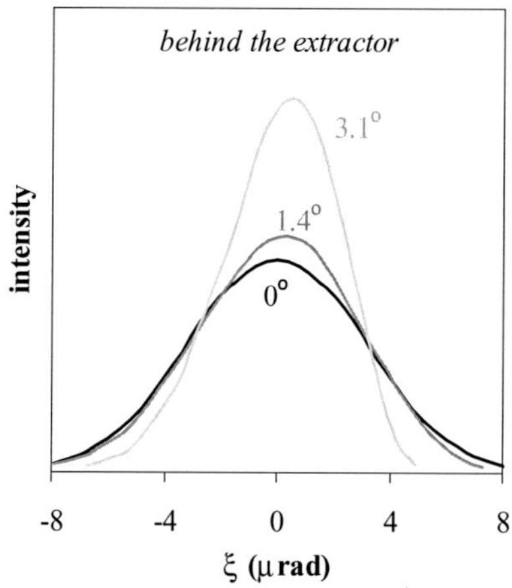

Figure 4.9 (1D) angular intensity distribution collected at different angles in the field-free zone behind extractor, at 2.5 mm from the emitter. For better comparison the distributions are aligned by shifting along the horizontal axis. T = 1,800 K; ϕ = 2.93 eV. (The angles 1.4° and 3.1° correspond to the final angles of cold electron launch positions of 50 nm and 100 nm from the facet center). *Abbreviation*: 1D, one-dimensional.

A motivation for this choice of definition will follow in the next section.

$E_{\tan}^{(FC=0.5)}$ can be expressed in terms of the mean tangential energy $\langle E_{\tan}\rangle$ from

$$0.5 = \frac{\int_0^{2\pi}\int_0^{\rho_{FC=0.5}} e^{-\left(\frac{\rho}{\rho_T}\right)^2} \rho d\rho d\theta}{\int_0^{2\pi}\int_0^{\infty} e^{-\left(\frac{\rho}{\rho_T}\right)^2} \rho d\rho d\theta} = \frac{\int_0^{E_{\tan}^{(FC=0.5)}} e^{\frac{-E_{\tan}}{\langle E_{\tan}\rangle}} dE_{\tan}}{\int_0^{\infty} e^{-\frac{E_{\tan}}{\langle E_{\tan}\rangle}} dE_{\tan}} = 1 - e^{\frac{-E_{\tan}^{(FC=0.5)}}{\langle E_{\tan}\rangle}}, \tag{4.5}$$

for which we have used the proportionalities $\rho^2 \propto E_{\text{tan}}$ and $\rho_T^{\;2} \propto \langle E_{\text{tan}} \rangle$. This gives

$$E_{\text{tan}}^{(FC=0.5)} = \langle E_{\text{tan}} \rangle \ln 2 . \tag{4.6}$$

Combining Eqs. (4.4) and (4.6), the virtual source size d_v containing 50% of the current is

$$d_v = 2\rho_{FC=0.5} = 2\sqrt{\ln(2)}\frac{r}{\alpha}\sqrt{\frac{\langle E_{\text{tan}} \rangle}{eV_{\text{ext}}}} = 1.67\frac{r}{\alpha}\sqrt{\frac{k_B T}{eV_{\text{ext}}}} . \tag{4.7}$$

The virtual source size decreases with increasing extraction voltage and decreasing temperature. (Note that if the extraction voltage is changed, but the suppressor voltage is kept constant, also the r/α lens effect factor changes—e.g., Fig. 3.20). The dependence of the virtual source size on the temperature and extraction voltage for a constant suppressor voltage is shown in Fig. 4.10.

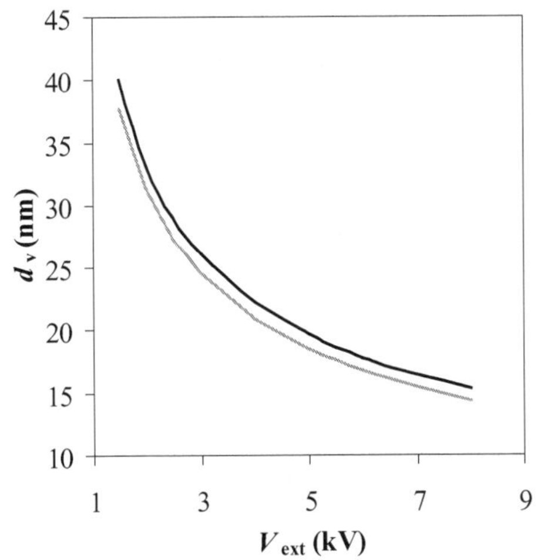

Figure 4.10 Virtual source size as a function of extraction voltage for a constant suppressor voltage, for 1,800 K (black) and 1,600 K (gray). Emitter geometry is as in Fig. 3.1.

For multibeam purposes, the optical design of the system will not only image the virtual source of the central beamlet but also image that of off-axis beamlets. Figure 4.2 has already shown that the virtual source position for off-axis beamlets is slightly different from that of the on-axis beamlet. Generally the size is also different. This is because electrons in the off-axis beamlet originate predominantly from a different area on the facet and experience a different lens effect than the beamlet from the facet center. The virtual source for off-axis beamlets will also be slightly asymmetric. This can already be seen from the angular intensity distributions in Fig. 4.9. In Fig. 4.11 we give the spatial intensity distributions associated with Fig. 4.9 for three different planes behind the facet (the virtual source plane of the central beamlet and planes at +0.5 μm and –0.5 μm).

It is noted that the intensity distributions are constructed from emission from a "line" in a rotationally symmetric simulation and can thus not be used to find the exact FW50 size of the virtual source for the off-axis beamlets.

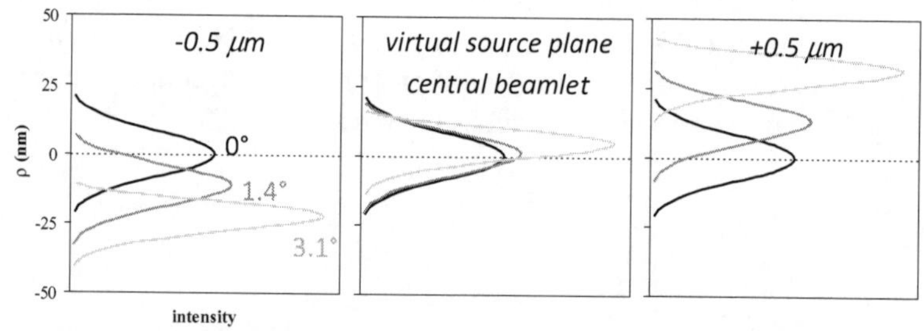

Figure 4.11 (1D) intensity distributions in three different planes behind the facet perpendicular to the optical axis, seen by three different collection angles in the field-free zone behind the extractor.

The size of the virtual source for these beamlets can be found, to first approximation, by assuming the field on the relevant facet area equals the field strength at the launch position of the cold electrons in the beamlet and by assuming an aberration-free lens effect between the relevant facet area and the aperture behind the extractor with a strength that equals that experienced by the cold electrons in the beam. To first approximation the virtual source size of an off-axis beamlet can then be found from

$$d_{\mathrm{v}} = 1.67 \frac{dr}{d\alpha} \sqrt{\frac{k_{\mathrm{B}} T}{e V_{\mathrm{ext}}}} . \tag{4.8}$$

An example is given in Fig. 4.12. As can be seen, for this approximation the virtual source size goes to zero for the outermost beamlets (see, e.g., Fig. 3.3: $dr/d\alpha \rightarrow 0$), which is associated with emission from near the facet edge.

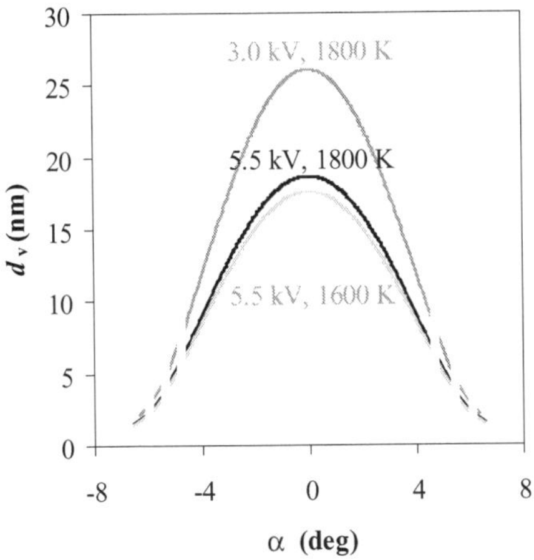

Figure 4.12 Virtual source size estimate as a function of the angle of the beamlet in the field-free zone. Emitter geometry in Fig. 3.1 in the standard gun geometry with a suppressor–extractor distance of 0.75 mm and a protrusion of 242 μm and with a suppressor voltage of −0.3 kV with respect to the emitter.

In summary, for a Schottky emitter, the position and size of the virtual source that is imaged in an optical system depend on the exact emitter and gun geometry, the operating conditions, and on which part of the facet beam will be used for the application.

For a Schottky emitter in normal operating conditions the virtual source of a beamlet is around 30 μm behind the facet. For on-axis beamlets it is on the optical axis and for off-axis beamlets up to tens of nanometers above or below the optical axis.

The size of the virtual source is typically a few tens of nanometers and will increase with increasing temperature and (for a constant field form) decreasing extraction voltage.

For the effect of the geometry of the emitter itself on the virtual source size, we can use some of the results of Chapter 3: if operated at the same extraction voltage, the virtual source size will be larger for an emitter with a smaller facet (but same tip size), with a larger tip (but same tip geometry), and with a larger cone angle in the tip vicinity (but equivalent tip end). It is noted that if one would compare emitters not for constant voltage, but for constant field, that the extraction voltage would have to be increased for the larger-facet emitter, the larger-tip-size emitter, and the emitter with the smaller cone angle, and this will reduce the difference in virtual source size a little bit. The difference in virtual source size for a constant field means that, for example, for the same source image size on the target a different magnification is required from the electron-optical system.

For off-axis beamlets the size of the virtual source is generally smaller than for the central beamlet and increasingly asymmetric for increasing angle of the beamlet with the optical axis of the system.

The effect of spherical aberrations of the facet extractor lens increases with increasing beam angle but is for the geometry investigated still negligible up to ~5 mrad half opening angle for beamlets that make a large angle with the optical axis. For the central beamlet the effect of the spherical aberration is for the given geometry negligible for beamlets up to at least ~20 mrad half opening angle.

4.2 Current in the Source Image: Practical Brightness*

When the virtual source position is known, the lens settings for the optical column can be chosen such that the beam is focused onto the sample to make an image of the virtual source scanning (transmission) electron microscopy (S(T)EM like) or is more collimated (TEM like). For any imaging or analysis of the sample a response signal is required: the detector needs to collect enough secondary, back-scattered, or transmitted electrons within a reasonable time step. The measured response signal depends on the incident current: how much current is in the virtual source image or in the coherent beam?

Probe size and current (similarly beam coherence and current) are interrelated: for example, for high-resolution work a small probe is desirable, but the probe cannot be too small, because the number of electrons or ions in the probe must be sufficient to actually see something. Currently, however, it is not possible to accurately calculate the current in a probe, because the probe current–probe size relation is ill-defined. Important to the amount of current in the probe at the target is how many particles the source can emit per unit area per solid angle per second (particle or charge flux density per unit solid angle). In electron- and ion-optics this quantity is the "richtstrahlwert" or "brightness" in A/m^2sr (Hawkes & Kasper,

*Most of this paragraph has been previously published and was reproduced with permission from *J. Vac. Sci. Technol. B*, **26** (2008) 949, Copyright 2008, American Vacuum Society (see also page 249 of this book).

1996a). The analogy in radiometry is "brightness" or "radiance" in W/m^2sr.

Because the beam brightness changes when the beam enters a section of different potential (in radiometry a medium of different index of refraction, n), the beam brightness at the source doesn't say much about the amount of current in the probe at the target. Nor does the beam brightness at the target say anything about the quality of the source. Calling the beam brightness at the target the "source brightness" (Müllerová, 1999) or "gun brightness" (Joy, 2005) or "brightness of the cathode" (Neo *et al.*, 2006) is confusing.

The relevant source property for probe-forming systems is "richtstrahlwert per volt" (Fink & Schumacher, 1975) or "normalized" or "reduced" brightness (Hawkes & Kasper, 1996a), which is the brightness divided by the beam potential, in A/m^2srV. The reduced brightness for infinitely small area and solid angle is the differential reduced brightness, which is directly proportional to the total number of electrons emitted with zero tangential energy (Shimoyama & Maruse, 1984). This source property is conserved throughout the column (Liouville's theorem, analogy in radiometry: conservation of brightness divided by n^2) as long as there are no interactions between the particles in the beam (see Section 4.4), and is the bridge between source performance and the amount of current in the probe.

4.2.1 The Definition of Practical Brightness

In general the differential reduced brightness B_{diff} is a function of the position in the plane, $\vec{r}$, and of the beam direction with respect to the normal of the plane, $\vec{t}$, the generalized brightness function in Hawkes and Kasper (1996a). The current in the specimen plane, I_p, can thus be calculated from

$$I_p = V_p \int_\Omega \int_A B_{\text{diff}}\left(\vec{r},\vec{t}\right) d\vec{r} d\vec{t}, \tag{4.9}$$

where V_p is the beam energy at the specimen and the integration is to be performed across the total area of the beam intensity distribution in the specimen plane and for all angles that arrive at the specimen.

The beam current that hits the specimen is determined by a current-limiting aperture. Usually this aperture cuts out a very small portion of the total emission cone, which allows for the approximation of uniform current density in the aperture plane and uniform angular intensity in the (virtual) source plane and the source image plane. For the source image plane this simplifies Eq. (4.9) to

$$I_{\mathrm{p}} = \Omega_{\mathrm{p}} V_{\mathrm{p}} \int_{\mathrm{image}} B_{\mathrm{diff}}(\vec{r})\, d\vec{r}\,, \tag{4.10}$$

which gives the current in the source image plane. Note that the integration is performed across the total area of the (virtual) source image, not the total area of the probe, since the probe also has contributions from diffraction and aberrations (see Section 4.3).

The Eq. (4.10) expression for the probe current does not yet contain the source image area A_{I}. For practical purposes we would like to have a relation such as

$$I_{\mathrm{p}} = \Omega_{\mathrm{p}} V_{\mathrm{p}} A_{\mathrm{I}} B_{\mathrm{pract}}\,. \tag{4.11}$$

This requires a definition of B_{pract} as

$$B_{\mathrm{pract}} = \frac{\int_A B_{\mathrm{diff}}(\vec{r})\, d\vec{r}}{A_{\mathrm{pract}}}\,. \tag{4.12}$$

For the source image plane the integration in Eq. (4.12) would be over the total source image area, and A_{pract} would be a practical measure for the size of the source image: $A_{\mathrm{I}} = \pi(d_{\mathrm{I}}/2)^2$ with d_{I} a practical source image diameter measure. There are many choices for d_{I}: full width half maximum (FWHM), full width containing 50% of the current (FW50), root mean square (RMS), etc. By choosing a measure for d_{I} Eq. (4.12) becomes a "practical brightness," which can be used to really quantify the relation between probe current and source image size.

The source image is the (de)magnified image of the (virtual) source with diameter d_{v}: thus $d_{\mathrm{I}} = Md_{\mathrm{v}}$, with M the column magnification. The practical brightness is the same in both planes. By applying the definition of Eq. (4.12) to the plane of the virtual source the practical brightness can be written as

$$B_{\mathrm{pract}} = \left[\frac{\int_A B_{\mathrm{diff}}(\vec{r})\, d\vec{r}}{A_{\mathrm{pract}}}\right]_{\mathrm{source}} = \frac{4I'}{\pi d_{\mathrm{v}}^{2} V_{\mathrm{ext}}}, \tag{4.13}$$

with I' the angular intensity of the source, I/Ω, V_{ext} the extraction voltage, and d_v the diameter of the virtual source. Crewe (1997) has also recognized Eq. (4.13) as the only meaningful brightness with respect to probe-forming systems, although he does not give a definition of d_v.

To complete the practical brightness definition in Eq. (4.13) we have to choose a measure for diameter d_v. For that we look at measures for the total probe size d_p because it makes sense to use the same size measure for the virtual source as for the total probe that contains the image of that virtual source. Reimer (1998) has pointed out that a definition of electron probe size should be free of assumptions about the electron–probe profile. This excludes the often-used FWHM. Jansen (1990) explains why RMS is unsuitable. A measure that assumes no particular shape of the probe intensity profile is the diameter $d(FC)$ containing a *f*raction of the *c*urrent (FC). And usually this is sufficient information: not the exact profile of the probe but the number of electrons within a certain area is what matters.

For FC we choose a value of 0.5, which is a convenient choice for a number of reasons. One is the relation between probe size and a common measure for resolution in scanning electron microscopy (SEM): the knife edge. Edge detection is often used to indicate resolution. In that method the probe is swept over an edge and the current behind the edge or on the edge is detected while it rises from 0% to 100%. A common measure for resolution is the distance between the 25% and 75% levels of the current. The relation between the measured knife edge width and the diameter of the probe will, however, depend on the shape of the probe intensity distribution and the definition of probe diameter. Figure 4.13 shows for three different intensity distributions the ratio of the probe diameter and the knife edge width, $d(FC)/d_{25\text{-}75}$, for different definitions of the diameter as defined by FC. As anticipated this ratio is a function of FC and also depends on the particular probe shape. There is, however, an intersection of the three curves very close to $FC = 0.5$. This means that (only) for $FC = 0.5$ one can conveniently convert between edge resolution and probe size without having to know the shape of the probe: $d_{FW50} = 1.76 d_{25\text{-}75}$. The inset in Fig. 4.13b shows d_{FW50} and $d_{25\text{-}75}$ for a top-hat intensity distribution.

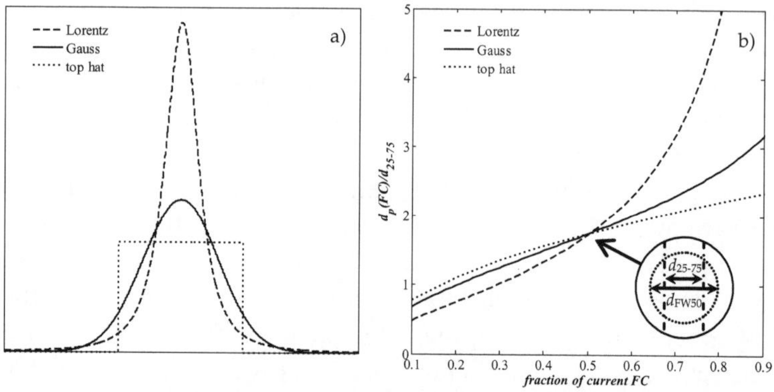

Figure 4.13 (a) Cross section through the probe, normalized for content, for a Gaussian, a Lorentzian, and a top-hat distribution (2D). (b) Ratio of the probe size $d_p(FC)$ and the 25–75% edge resolution, d_{25-75} for different current fractions. The inset shows d_{FW50} and d_{25-75} for a top-hat intensity distribution. *Abbreviation*: 2D, two-dimensional.

Two additional arguments follow from the point that it makes sense to use the same size measure not only for the virtual source and the total probe but also for all contributions to the total probe. In that case it is convenient that the FW50 of the chromatic aberration contribution, contrary to, for example, the FWHM of this contribution, is approximately independent of the shape of the energy distribution (Barth & Nykerk, 1999). Also, for electron probes, the inevitable diffraction contribution (the intensity distribution produced by Fraunhofer diffraction around a circular aperture: airy pattern) has an FW50 width that is very close to its FWHM (4% difference).

Finally, the probe intensity profiles encountered in practice are often approximately Gaussian, for which the FW50 equals the FWHM. This should facilitate adoption of the FW50.

Thus with d_v in Eq. (4.13) taken to be the FW50 value the practical brightness definition is complete and we can drop the subscript "pract."

If the practical brightness at the extractor (Eq. 4.13) is conserved down to the target (see Section 4.4), we can quantify the relation between probe current and source image size (Eq. (4.11)):

$$I_p = \pi\alpha_p^2 V_p \frac{\pi}{4} d_I^2 B, \tag{4.14}$$

with α_p the half opening angle of the beam at the target and B determined by the source properties, Eq. (4.13).

For TEM purposes one would like to quantify the amount of current in a spatially fully coherent beam. The coherence length can be given as

$$X_{\text{coh}} = \frac{\lambda}{2\theta_c}, \tag{4.15}$$

with λ the electron wavelength and θ_c the FW50/2 of the angular intensity distribution that is collected at each point on a uniformly illuminated target (or the half-angle from the target to the FW50 source image). This gives for the amount of current in a fully coherent beam

$$I_{\text{coh}} = \pi\theta_c^2 V_b \frac{\pi}{4} X_{\text{coh}}^2 B = \pi V_b \frac{\pi}{4} \frac{h^2}{8meV_b} B = 0.93 \cdot 10^{-18} B, \tag{4.16}$$

with B again determined by the source properties, Eq. (4.13).

Because the practical brightness is the only brightness that can actually quantify the amount of current in a probe, we hope that for future characterizations of probe-forming systems the practical brightness will replace the use of all other brightness definitions.

4.2.2 How to Get the Practical Brightness of a Source?

The practical brightness of a source at a given extraction voltage can be evaluated when its angular intensity and FW50 virtual source diameter are known, Eq. (4.13).

The FW50 virtual source diameter can be calculated from the shape and size of the intensity profile of the virtual source (see also Section 4.1). For some source types this is known: thermionic emitters, for example, have a Gaussian source intensity profile with a temperature-dependent width (Hawkes & Kasper, 1996b). For other source types only their shape is known, such as for liquid metal ion sources (which have a shape with much stronger tails than the Gaussian (Ward *et al.*, 1988), or both shape and size are unknown, such as for the nanotube (de Jonge, 2004) or the nanotip (Vladár *et al.*, 2006) emitter. In that case the practical brightness of the source can only be found by measuring the source intensity profile.

There is an ongoing effort to measure the size and shape of probes in general. The most commonly used technique for spot size measurement in probe-forming systems is the knife edge scan. This method generally works for probe sizes down to a few nanometers (Rishton, 1984; Kratschmer *et al.*, 1988), although it has potential sources of error (Babin, 2006). A different method is using a point projection microscope (de Jonge, 2004; Fransen, 1996), but to extract a size from the experimental Fresnel fringes patterns one has to assume the shape of the source intensity distribution.

Also Fourier methods have been applied (Joy, 2002), but a routine that can deal with unknown probe shapes is still under development (Babin, 2006; Carasso, 2002; Liddle, 2004). Perhaps the most straightforward method to obtain the size and shape of a virtual electron source is to put the source as emitter in a transmission electron microscope and use the column to make a highly magnified image of the source (Fransen *et al.*, 1999).

Once the probe size is known, it still includes contributions from diffraction and aberrations, in addition to the desired source image intensity profile. But when the latter dominates the probe, the total size can be corrected with the root power sum (RPS) method (Section 4.3) to yield the size of the virtual source.

4.2.3 The Intrinsic Practical Brightness for Thermionic, Schottky, and Cold Field Emission Electron Sources

The practical brightness of thermionic, Schottky, and cold field electron emitters is a special case because it can be shown to be directly proportional to the current density at the emitting surface, which follows from the electron emission theory in Chapter 1 when the local work function, temperature, and field are known. This is only when for these emitters the effect of electron–electron interactions (Section 4.4) between the source and the extractor on the virtual source size is negligible, but this often is a safe assumption.

Using Eq. (3.5) for the angular intensity and Eq. (4.7) for the virtual source size, Eq. (4.13) gives for the practical brightness of thermionic and Schottky sources

$$B = \frac{1}{\ln(2)} \frac{ej}{\pi k_{\mathrm{B}} T} = 1.44 \frac{ej}{\pi k_{\mathrm{B}} T}, \tag{4.17}$$

and for the practical brightness of cold field emitters (where d and not $k_{\mathrm{B}}T$ is the mean tangential energy, see Section 1.4.3):

$$B = 1.44\frac{ej}{\pi d}. \tag{4.18}$$

Note that the lens effect is no longer required for the practical brightness: the practical brightness of these sources follows from the electron emission theory for infinite planes.

But if the relation between the current density on the emitting surface and the angular intensity at the extractor is known, Eq. (3.5), only the angular intensity is required to verify the practical brightness experimentally:

$$B = 1.44\frac{eI'}{\pi k_B T}\left(\frac{\alpha}{r}\right)^2. \tag{4.19}$$

The factor α/r—or the equivalent $1/f$ (Fujita & Shimoyama, 2005a, 2005b, 2006) or m_α/r_{tip} (Swanson & Schwind, 2008)—can be found from SEM imaging and the ray tracing simulations as demonstrated in Chapter 3.

The factor 1.44 in the brightness equations is a new result. Note that the quantities $ej/(\pi k_B T)$ and $ej/(\pi d)$ in Eqs. (4.17) and (4.18) are the often quoted axial differential reduced brightness values $B_{diff}(0)$ of, respectively, thermionic and Schottky sources and cold field emitters (Hawkes & Kasper, 1996a). The differential axial reduced brightness $B_{diff}(0)$ is the current from an infinitesimal area at the center of the source intensity profile into an infinitesimal solid angle. Remember that when electron sources are used for probe formation, the differential axial reduced brightness is not the relevant parameter, because in probe formation the whole of the virtual source is imaged. The practical brightness is proportional to the differential axial reduced brightness, but the proportionality factor depends on the shape of the source intensity profile. $B_{diff}(0)$ can therefore not replace B to calculate the probe current with Eq. (4.14), as is done in (Fujita & Shimoyama, 2005b), nor should FW50 brightness measurements be compared to the differential values, as is done in (Fransen *et al.*, 1999). It is noted that the practical brightness can be increased (at the cost of beam current) by imaging the source onto a beam-limiting aperture that cuts off the tails of the intensity distribution. This method can increase the practical brightness up to a maximum of $2ej/(\pi k_B T)$, which is associated with the situation in which the profile cut from the Gaussian intensity distribution can be approximated by a top-hat distribution with a constant differential reduced brightness of $B_{diff}(0)$.

4.3 Total Probe Size: Source Image Plus Diffraction Plus Aberration Contributions

With the practical brightness of the source known, Eq. (4.14) gives the desired relation between probe current and source image size. But the probe in a system is larger than the source image because of contributions from aberrations and diffraction. To get the total probe size, aberration and diffraction contributions have to be added to the source image size. Too often a quadrature addition of inconsistently defined diameters is used, with each diameter representing one of the contributions (addition of, e.g., an FWHM for the source image and an FW100 for spherical aberration).

We advocate the use of a consistently defined diameter for all contributions and discourage the use of the quadrature addition. A common mistake for electron probes is the quadrature addition of (consistently defined) FWHMs of the source image and the diffraction-limited spot, with the argument that both are (approximately) Gaussian. This would be valid if the source image and the diffraction-limited spot are independent distributions that can simply be multiplied to yield the total probe distribution, but this is not so: the total probe is a convolution of the source image and the point-source image that contains the blur due to diffraction and aberrations. Using only FW50 diameters the so-called RPS method fits the full simulations (Barth & Kruit, 1996) better than the quadrature addition. Following the RPS method the minimum FW50 total probe size at the target is

$$d_p = \sqrt{\left((d_A^4 + d_S^4)^{1.3/4} + d_I^{1.3}\right)^{2/1.3} + d_C^2} \,. \tag{4.20}$$

Each of the individual terms is the exact FW50 value that would exist in the absence of the other three. The powers are those that give a good approximation to complete simulations of electron or ion probes. The FW50 source image diameter d_I follows from Eq. (4.14), and the other diameters represent the diffraction contribution, d_A, and the contributions from chromatic and spherical aberration, d_C and d_S, respectively:

$$d_A = \frac{0.66 \cdot 10^{-9}}{\sqrt{V_p}\,\alpha_p}, \tag{4.21}$$

$$d_C = 0.6 C_C \frac{\Delta E_{FW50}}{V_p} \alpha_p , \tag{4.22}$$

$$d_S = 0.18 C_S \alpha_p^3 . \tag{4.23}$$

C_C and C_S are the chromatic and spherical aberration coefficients of the system. When the last probe-forming lens highly demagnifies the aberrations of preceding lenses, as is usually the case, it suffices to use the coefficients of the last lens. ΔE_{FW50} in Eq. (4.22) is the FW50 of the energy distribution of the electrons (see Section 1.4.3).

Because of the opposite dependence of d_I and d_A versus d_C and d_S on α_p, there is for each desired combination of probe current and beam energy a different optimum angle that minimizes the total probe size. Figure 4.14 shows an example of the optimum probe current–probe size relation for an imaginary scanning electron microscope and a beam energy of 30 kV.

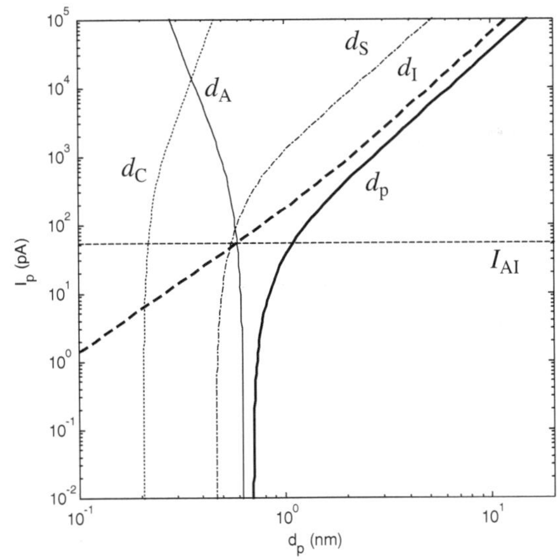

Figure 4.14 Typical probe size–probe current relation calculated with Eqs. (4.14) and (4.20)–(4.23). Input parameters: $B = 5 \times 10^7$ A/m^{2}srV, $\Delta E_{FW50} = 0.6$ eV, $V_p = 30$ kV, $C_S = 12$ mm, and $C_C = 4.4$ mm.

For each point on the d_p curve in Fig. 4.14 the beam half opening angle at the target was optimized (here from ~6 mrad to ~14 mrad). Probe currents much less than 1 pA are not practical, but we include these to illustrate how fast the initial rise of I_p with d_p is. We did not include probe currents greater than 20 nA because in this example we have neglected the gun aberrations, which become important at larger probe currents, when the acceptance angle at the gun has to become large (Kruit *et al.*, 2006).

As can be seen from Fig. 4.14 the probe current for which the contributions from diffraction and the source image have the same size ($d_A = d_I$) is a useful characteristic: it is where the probe size starts increasing with increasing probe current. Because d_I and d_A have the same beam energy and angle dependence this probe current I_{AI} is determined *solely* by the practical brightness of the source: $I_{AI} = 1.08 \times 10^{-18}B$. This "brightness" current I_{AI} separates the different regimes that can be distinguished in Fig. 4.14: for $I_p < I_{AI}$ the contributions d_A and d_I balance d_S, while for $I_p > I_{AI}$, d_A becomes negligible and the balance is between d_I and d_S. Note that at lower beam energy d_c will play a more important role.

In general, each section of the $I_p(d_p)$ curve has its own figure of merit for the source and for the column properties, which can be found from the equations. Increasing the right figure of merit will improve the system performance for the curve section of interest. Appendix B gives a scheme to determine which probe size contributions are dominant, which is the appropriate figure of merit for the source and the column, and, for a desired probe current, what the minimum probe size is together with the optimum half opening angle of the beam and optimum (de)magnification of the virtual source. A requirement is that the aberration coefficients of the final lens be known (and for large probe currents and/or sources with low angular current densities, the aberration coefficients of the gun lens). The scheme could help users to better understand their system.

In summary, the practical brightness and energy spread of the beam in combination with the RPS method allow us to quantify the relation between probe current and probe size for a probe-forming system and find the performance-limiting factors for each application.

To improve performance the brightness and energy spread can be changed by changing the operating parameters of the source. So far we have neglected the effect of interactions between the electrons in the beam, allowing calculation of the practical brightness and energy spread with the theory in Chapter 1 and in Section 4.2.3.

These properties we call the intrinsic properties. Their relation is shown in Fig. 4.15 as a function of the field and temperature at the surface.

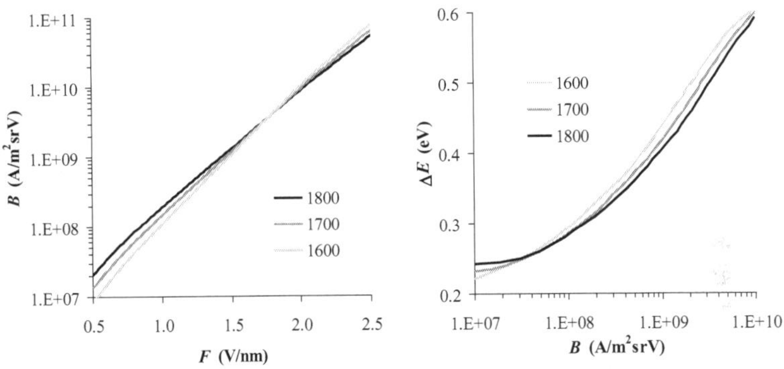

Figure 4.15 Effect of field strength and temperature (in K) on practical brightness and the relation between intrinsic brightness and intrinsic energy spread (FW50). The work function is temperature dependent and 2.93 eV at 1,800 K.

The brightness and energy spread of the beam at the target are, however, only at best equivalent to the intrinsic values: as soon as electrons leave the emitter they can start interacting with other electrons in the beam and change the brightness and energy spread of the beam. This can affect the probe current–probe size (coherence) relation at the target.

The effect of Coulomb interactions in the first few millimeters of the beam path is the topic of the next section.

4.4 The Effect of Electron–Electron Interactions in the Beam

Generally the effect of longitudinal interactions between electrons in the beam is distinguished from the effect of lateral interactions. The change in axial velocity resulting from the interactions causes broadening of the energy distribution and is called the Boersch effect (Boersch, 1954). The change in longitudinal velocity is called trajectory displacement and causes brightness loss through an enlargement of the virtual source.

How much the intrinsic beam properties change between the emission site and the final target depends on the positions and velocities of the electrons with respect to each other along their path: for example, a crossover of a large current beam of slow electrons will generate more particle interactions than a crossover of a low current beam with fast electrons.

Because of the statistical nature of the electron–electron interactions one cannot correct for it. It is therefore valuable to know in what way the Boersch effect and trajectory displacement are affected by the operating conditions and parameters such as the geometry of the emitter, gun and, column, as it will aid in finding the optimal design and operation.

The energy and virtual source size broadening gained in the column can be minimized by tuning the column design (Jiang & Kruit, 1996s). Limiting the total beam current early in the column is a very effective way to do so. To quantify the effect of the interactions in the column, adequate analytical approximations (Kruit & Jansen, 2008) as well as Monte Carlo simulation results are available (Jansen, 1990; Read & Bowring, 2003; Read & Bowring, 2004; http://www.mebs.co.uk*). The broadening in the gun cannot so easily be minimized: the first position to limit the beam current is at the extractor, but the damage may already be done because close to the emitting surface the electron density is high and the velocities are low. Simulations for the gun region are complicated. A full calculation of the effect of Coulomb interactions would require the calculation of the number of electrons, and the energy distribution of the electrons, emitted per unit time by each point on the emitter surface, following by detailed ray tracing of the emitted particles, taking into account both the (rapidly varying) field and the interaction between the particles (Radlicka & Lencová, 2008). Such an approach is currently still rather time consuming and has not been applied to the case of a real Schottky emitter gun yet. (There is one account of a Monte Carlo approach to get the Boersch effect for Schottky emitters, but it does not provide much detail [Thomson, 1994].)

The effect of Coulomb interactions on the brightness or energy spread of Schottky emitter guns has been addressed before (Gesley, 1991; Fransen *et al.*, 1998; Van Veen *et al.*, 2001; Sakawa *et al.*,

---

*E. Munro, SOURCE/BOERSCH, Munro's Electron Beam Software Ltd., http://www.mebs.co.uk.

2003; Kim *et al.*, 1997a; Thomson, 1994), but generally either the brightness or the energy spread is investigated, not both. As shown above you generally need both to characterize the performance of a system. For Schottky emitters both theoretical results as well as experiment-based results have been presented. A first difficulty that applies to either approach is that you are required to know how the operating conditions (extraction voltage, heating current) relate to the brightness and energy spread of the beam that leaves the emitter surface. This has been addressed in Chapter 3 and in the preceding sections of this chapter. Then, for experimental methods the difficulty is how to correctly subtract the calculated intrinsic energy spread or brightness from the measured total energy distribution or brightness. In theoretical approaches adequate models have been sought to determine the effect of Coulomb interactions (Knauer, 1981; Jansen, 1990; Modinos & Xanthakis, 1991) without the need for full-particle tracing, raising the question of how to correctly add the contribution from Coulomb interactions to the intrinsic properties.

The literature shows different models for the different steps involved (intrinsic properties as a function of operation parameters, the effect of Coulomb interactions, adding/subtracting contributions), making it difficult to compare the results. Although quite a few aspects have been addressed before separately, the total picture of the effects of choices in extraction voltage, heating current, and emitter size on the brightness *and* its associated energy spread is currently lacking.

In this paragraph we will focus on the effect of the interactions on the brightness and energy spread in these first few millimeters of the beam path. Below, in Section 4.4.1, we give a theoretical approach based on the theory in Chapter 1, charge density method simulations, and for the effect of the interactions, the equations derived by Jansen (1990) and Kruit and Jansen (2008). In Section 4.1.2 we present an experimental approach to determine the effect of the interactions on the basis of energy spread data.

4.4.1 Simulations

Below we will first address the general equations that are used to determine the effect of Coulomb interactions and then apply these to the case of four different Schottky emitters. The final paragraph discusses how to add the effect of the interactions to the intrinsic beam properties and find the practical brightness and energy spread of the beam as a function of the distance from the emitting surface.

4.4.1.1 General equations

The effect of trajectory displacement and the Boersch effect will be calculated on the basis of the analytical equations of Guus Jansen (1990) and Kruit and Jansen (2008). Jansen's equations are based on the so-called extended two-particle approximation. In this approach it is assumed that the positions in the phase space of all field particles are statistically independent and that the displacement experienced by a test particle due to N field particles can be expressed as a sum of all two-particle effects. Jansen has considered beam segments at constant potential with both uniform and Gaussian (angular) current density distributions. It was found that a variety of regimes can be distinguished, depending on the particle density as well as on the geometry of the beam, where each regime can be characterized by a particular shape of the displacement distribution: Gaussian, Holtsmarkian, Lorentzian, or a pencil beam. Gaussian behavior occurs for large particle densities only. For lower particle densities, the interaction distributions show distinct non-Gaussian features such as a narrow core and long tails.

For each of the regimes Jansen has fitted his results to provide analytical and thus more practical equations for the FW50 of the displacement as a function of the beam parameters (Kruit & Jansen, 2008). An interpolation equation to add up the results of the different regimes is also given so that it is not necessary to know a priori which regime applies.

To calculate the displacement for beams with fast-changing potentials, such as for the beam between the emitter and the extractor, the slice method can be used (Jiang, 1996b). In this method the beam is sliced into segments for which the potential is assumed to be constant. The total displacement can then be found by calculating the displacement in each slice individually with Jansen's equations and adding up all slice contributions.

The slice method can be applied if the beam characteristics give rise to displacement distributions of the so-called "pencil beam" or "Holtsmark" type (Kruit & Jansen, 2008). In a pencil beam the lateral dimension of the beam is small compared to the separation of the particles. For a Holtsmarkian displacement distribution the lateral dimension of the beam is large relative to the separation of the particles.

4.4.1.1.1 *Calculation of the trajectory displacement*

Kruit and Jansen (2008) have given equations for the limiting cases of the pure pencil beam regime and the pure Holtsmark regime for cylindrical beam segments with a uniform current density distribution. For beam segments of constant beam potential and the z-direction along the optical axis, the FW50 angular deflection per meter, φ, due to the interactions is for pencil beam and Holtsmark type, respectively (Kruit & Jansen, 2008):

$$\varphi_{\mathrm{pb}} = 0.429 \frac{m^{3/2}}{\varepsilon_0 e^{7/2}} \frac{I^3 r(z)}{V(z)^{5/2}} \tag{4.24}$$

and

$$\varphi_{\mathrm{H}} = 0.128 \frac{m^{1/3}}{\varepsilon_0} \frac{I^{2/3}}{V(z)^{4/3} r(z)^{4/3}}, \tag{4.25}$$

with I the beam current, $r(z)$ the beam envelope, and $V(z)$ the beam potential. Jiang (1996b) has given an equation that covers both regimes:

$$\varphi = \left(\frac{T_1 D_\lambda^{18/7} D_\mathrm{r}^6 I^{18/7} r(z)^{6/7} V(z)^{-15/7}}{T_4 + T_2^{1/7} D_\mathrm{r}^6 D_\lambda^2 I^2 r(z)^2 V(z)^{-1}} \right)^{7/6}, \tag{4.26}$$

with $D_\lambda = \sqrt{m}/(\pi 2^{7/2} \varepsilon_0 \sqrt{e})$, $D_\mathrm{r} = (2\pi\varepsilon_0/e)^{1/3}$, and for a cylindrical slice $T_1 = 4.618 \times 10^{-2}$, $T_2 = 2.041 \times 10^5$, and $T_4 = 6.25 \times 10^{-2}$.

According to Van Veen *et al.* (2001) the Holtsmark regime is the relevant regime for the trajectory displacement in beams from the end facet of a Schottky emitter.

The trajectory displacement per slice can be found by multiplying the angular deflection per meter with the distance between the slice and the point where the beam angle of the slice crosses the optical axis (its image plane). To calculate the contribution of a slice to the trajectory displacement in the virtual source plane, the trajectory displacement of the slice is divided by the magnification between the image plane of the slice and the virtual source plane. The total FW50 blur in the virtual source plane can be found by integration:

$$d_{\mathrm{tra}} = \int_z \frac{r(z)}{\alpha_{\mathrm{vs}}} \varphi \sqrt{\frac{V(z)}{V_{\mathrm{vs}}}} dz, \tag{4.27}$$

with α_{vs} and V_{vs} the beam angle and beam potential in the virtual source plane (equivalent to the beam angle and potential in the field-free zone behind the extractor, α_{ext} and V_{ext}). Equation (4.27) assumes the configuration of particles remains the same in the passage to the next beam segment (fully correlated).

4.4.1.1.2 *Calculation of the Boersch effect*

In the pencil beam and Holtsmark regime the FW50 broadening, in eV/m, in a cylindrical slice at constant potential with uniform current density can be found with

$$E_{\mathrm{pb}} = 0.642 \frac{m}{\varepsilon_0 e^2} \frac{I^2}{V(z) r(z)^{4/3}} \tag{4.28}$$

and

$$E_{\mathrm{H}} = 0.149 \frac{m^{1/3}}{\varepsilon_0 e^{1/2}} \frac{I^{2/3}}{V(z)^{1/3} r(z)^{4/3}}. \tag{4.29}$$

Also for the Boersch effect Jiang (1996b) has given an equation that covers both regimes:

$$E = \left(\frac{B_1 D_\lambda^3 D_\mathrm{r}^6 I^3 V(z)^{-3/2}}{1 + B_2^{1/4} D_\mathrm{r}^6 D_\lambda^2 I^2 r(z)^2 V(z)^{-1}} \right)^{2/3}, \tag{4.30}$$

with, for a cylindrical slice, $B_1 = 7.861 \times 10^{-2}$ and $B_2 = 6.828 \times 10^9$.

The total Boersch effect can be found by integrating over all segments up to the plane where the Boersch effect is to be evaluated, at z_1:

$$\Delta E_{\mathrm{Boersch}} = \int_0^{z_1} E dz \tag{4.31}$$

Fransen *et al.* (1999) have calculated the Boersch effect for a Schottky gun with a different approach, in which the potential and particle density variations in the beam between the emitter and the extractor are ignored completely. They have assumed the region is field free and that the emitted electrons gain all energy very close to the source. With this approximation the Boersch effect can be found with Jansen's equations for a beam segment a constant potential with a crossover. On the basis of these equations Fransen *et al.* (1999) have predicted for Schottky sources distributions between Lorentzian and Holtsmarkian. In support of this, Thomson (1994), who used Monte Carlo simulations, also found distributions with stronger tails than for a Gaussian.

In this "one segment" model, the emitting surface is assumed to be at the crossover in the segment, and to take into account that electrons can only interact for half of the beam segment (after they have been emitted), the original equations are divided by two. For the Lorentz and the Holtsmark regime the Boersch effect contributions are (with ΔE in eV) (Kruit & Jansen, 2008)

$$\Delta E_{\text{FW50L}} = 1.9 \times 10^5 \pi \frac{I'\alpha_{\text{ext}}}{V_{\text{ext}}^{1/2}}, \tag{4.32}$$

$$\Delta E_{\text{FW50H}} = 4.88\pi^{2/3} \frac{(I')^{2/3}\alpha_{\text{ext}}^{1/3}}{r_{\text{c}}^{1/3} V_{\text{ext}}^{1/3}}, \tag{4.33}$$

and these can be added to give the total Boersch effect using

$$\frac{1}{\Delta E_{\text{FW50}}} = \left(\frac{1}{\Delta E_{\text{FW50L}}^4} + \frac{1}{\Delta E_{\text{FW50H}}^4} \right)^{1/4}. \tag{4.34}$$

Below we compare the results of the slice method with the one segment model, Eqs. (4.32)–(4.34).

4.4.1.2 Application to Schottky emitters

4.4.1.2.1 *Assumptions*

To investigate the effect of Coulomb interactions for different emitter geometries, we will investigate the four different emitter geometries given in Fig. 3.27 in the standard gun geometry with a distance between the suppressor and the extractor of 0.75 mm, a protrusion of 0.25 mm, and a fixed suppressor voltage of –0.3 kV with respect to the emitter. The first few millimeters behind the extractor are assumed to be field free. The extractor potential is varied between 2 kV and 15 kV. The temperature is varied between 1,600 K and 1,900 K, with a work function of 2.93 eV at 1,800 K.

Figure 4.16 gives the beam potential and current density for the beam center (for 1,800 K) as a function of the distance from the emitting surface for a field at the facet center of 1 V/nm. This figure shows how fast the beam properties change and that the effect of Coulomb interactions can be expected to be different for the different tips: on average the beam potential and current density are higher for the larger emitter.

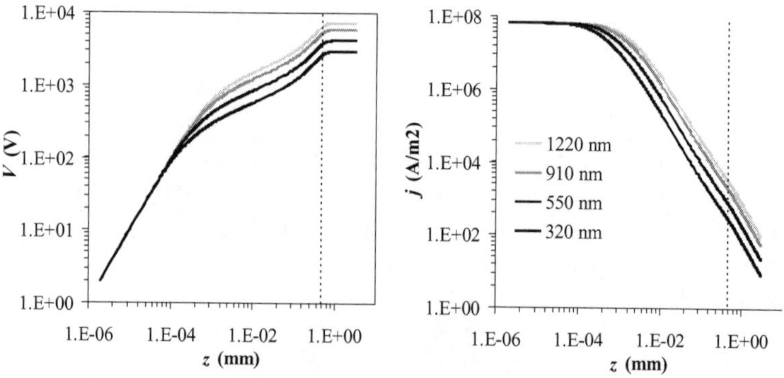

Figure 4.16 Beam potential and central current density as a function of distance from the emitting surface (at 1,800 K) for a field at the facet center of 1 V/nm. The dashed line indicates the position of the extractor.

In addition to the longitudinal changes in current density, it was shown in Chapter 3 that the current density in the beam is also changing laterally, on its way from the emitter to the extractor. It was also shown that the changes of the current density distribution in the beam are different for different emitters (Fig. 3.29). Equations (4.24) and (4.25) and Eqs. (4.28) and (4.29) apply to slices with uniform current density distributions, and similar equations for Gaussian current density distributions are also available, but we have shown that the facet beam of a Schottky emitter is generally not uniform nor Gaussian in current density.

The goal here is to find a good estimate for the total trajectory displacement for the center of the facet beam for the first few millimeters of the beam path. We now make the major assumption that can be done by neglecting the variations in current density on the facet and ignore the spherical aberration of the lens between the emitter and the extractor.

The validity of the assumption is connected to the question to what extent the electrons at the perimeter of the facet beam contribute to the displacement of the electrons in the center. Already it is noted that the central part of the current density profile is "flatter" for larger emitters, meaning that for larger emitters the effect of the different intensity at the beam edge is probably less relevant.

To investigate the assumption, Fig. 4.17 gives for the facet beam of one of the emitters the angular deflection as a function of the beam length for different initial beam diameters, calculated assuming a 100% pencil beam, Eq. (4.24); a 100% Holtsmarkian beam, Eq. (4.25); and the equation that covers both types, Eq. (4.26). For each beam it is assumed that the facet area outside the beam is not emitting, as if that part of the facet is covered with a high-work-function material.

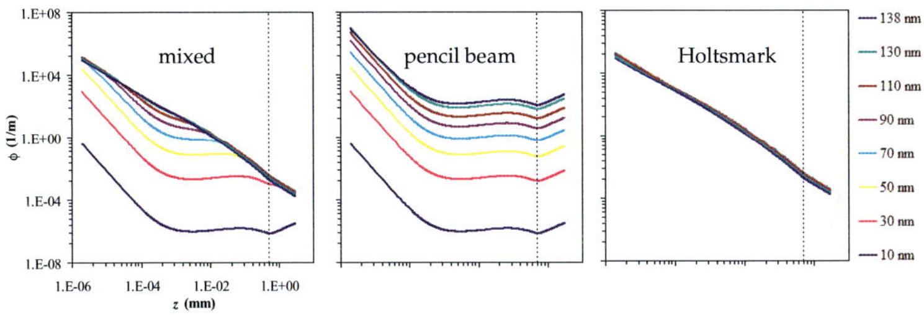

Figure 4.17 Angular deflection per meter as a function of distance from the facet along the optical axis for different trajectories calculated with Eqs. (4.24)–(4.26) for a field of 1 V/nm on the center of the facet of the 550 nm emitter (Fig. 3.27) at 1,800 K. The dashed line indicates the position of the extractor.

In the left plot of Fig. 4.17 it can be seen that φ increases with increasing initial beam size but becomes independent of the beam diameter beyond an initial beam radius of ~90 nm. This means that the electrons present outside this envelope do not increase the deflection any further. In other words, the deflection experienced by electrons in the center of the beam is determined by the local environment and is independent of the beam size. Because for the 90 nm beam radius the initial current density variations and lens aberrations are considered small enough, we conclude that the pure Holtsmark description is adequate for the case of Fig. 4.17. The best estimate for the displacement for the center of the beam is considered to be that calculated for a small beam envelope $r(z)$ (that spreads out the least) with the initial current density of the facet center. We will take the envelope that is bound by the ray path of an electron launched from the facet at 10 nm from the optical axis.

It remains to be seen whether the Holtsmark assumption is also true for other conditions. For example, for low current densities at the emitting surface the full facet beam may still be in the pencil beam regime. This is verified by comparing the beam diameter with the particle separation. The beam diameter is calculated from the trajectory of an electron launched from a position on the facet close to the edge, and the average axial distance between electrons is estimated from $e\sqrt{(2eV/m)}/I$, where the beam current I was calculated from the central current density and the facet diameter. For the operating conditions and emitters under investigation this analysis yields that for the lowest extraction voltages and the smallest emitter, the beam does start off in the pencil beam regime but enters the extended regime at a few (tens of) microns from the emitting surface. For these cases, assuming Holtsmark behavior will thus overestimate the effect of Coulomb interactions, but the effect of the interactions at such low current densities will be small anyhow. It is therefore chosen to calculate the effect of the Coulomb interactions for the beam center, assuming pure Holtsmark behavior for all geometries and conditions investigated, with an initial current density in the beam equal to the central current density on the facet.

4.4.1.2.2 *The contribution from Coulomb interactions*

Figure 4.18 gives the FW50 trajectory displacement and the FW50 Boersch effect as a function of the distance from the emitting facet for the 550 nm emitter of Fig. 3.27 for different extraction voltages and a constant temperature (left) and for different temperatures and a constant extraction voltage (right).

Figure 4.18 shows that the trajectory displacement is a strong function of z. The trajectory displacement will thus keep increasing unless the beam is apertured (Jiang, 1996b). On the contrary, the Boersch effect shows only a weak z dependence beyond the extractor. It is also clear that both the trajectory displacement and the Boersch effect are larger for a higher extraction voltage and a higher temperature.

Figure 4.19 compares the result for the four different emitters operating at 1,800 K and at an extraction voltage that gives 1 V/nm at the facet center.

Figure 4.19 shows that for constant temperature and field strength, the trajectory displacement and Boersch effect are larger for larger emitters. Because all four emitters have an equivalent intrinsic energy spread this also means that the largest emitter suffers the most from the Boersch effect. It is not so, however, that the largest emitter also suffers the most from the trajectory displacement, because the intrinsic virtual source size is not the same for each emitter (for the particular example 13 nm, 18 nm,

24 nm, and 28 nm for increasing tip size). To find the total energy spread and the new source size, the Boersch effect and trajectory displacement need to be added to the initial properties. This will be done in the next paragraph.

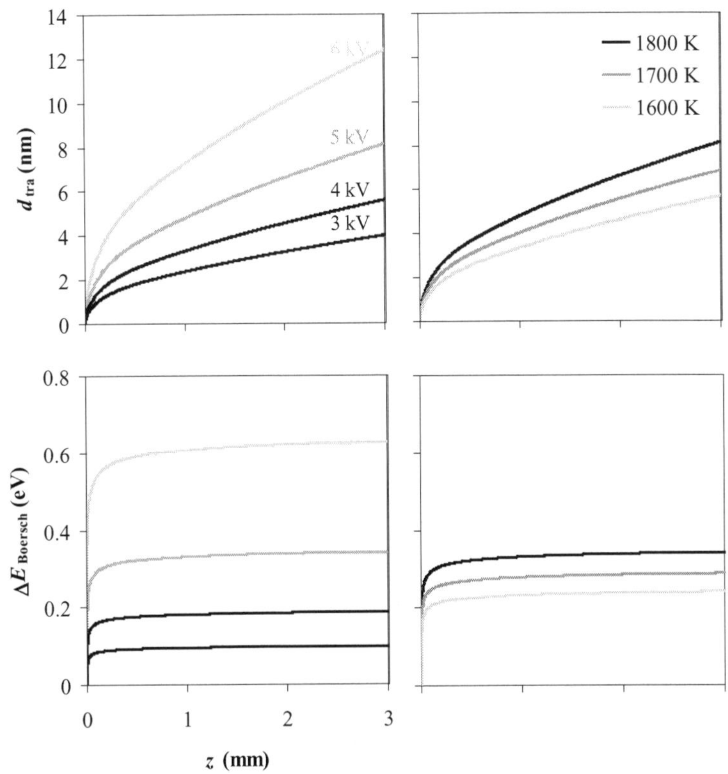

Figure 4.18 FW50 trajectory displacement (top) and FW50 Boersch effect (bottom) as a function of the distance from the emitting surface z, calculated with Eqs. (4.25), (4.27), and (4.29), for the 550 nm emitter operated at different extraction voltages (left) at a temperature of 1,800 K and at different temperatures at an extraction voltage of 5 kV (right).

Figure 4.19 also gives the Boersch effect calculated with Eqs. (4.32)–(4.34), which ignore the potential gradient. For the half opening angle α_{ext} we take the half opening angle of the facet beam, which is ~7º, and we take the facet radius as the crossover radius, r_c. For comparison the markers have been placed at 3 mm, although the results are independent of the beam length. It can be seen that for larger emitters the two models yield approximately the same result, but for smaller emitters the slice method result is higher. The slice method is indeed expected to overestimate the true Boersch effect due to the Holtsmark regime assumption, which is less applicable to the smaller emitters in Fig. 4.19 because of the lower current density in the beam.

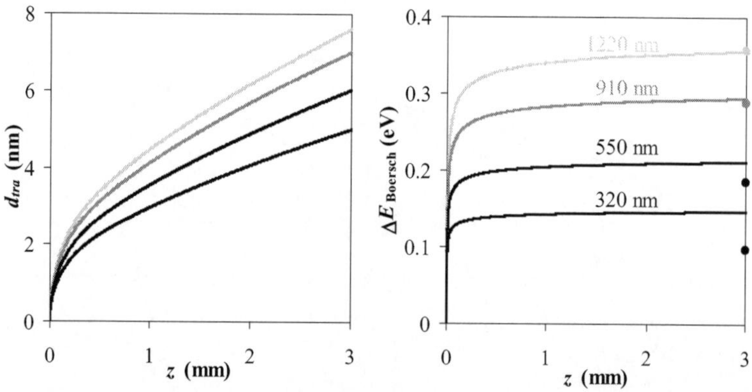

Figure 4.19 FW50 trajectory displacement and FW50 Boersch effect as a function of the distance from the emitting surface z for different emitter geometries, calculated with Eqs. (4.25), (4.27), and (4.29). The markers in the Boersch plot are calculated with Eqs. (4.32)–(4.34). The field strength at the facet center for all emitters is 1 V/nm and the temperature 1,800 K.

4.4.1.3 Adding contributions together

To add the effect of the Coulomb interactions to the intrinsic properties we need to add distributions that generally have a different shape: the intensity distribution in the virtual source is Gaussian, and the intrinsic energy distribution is often something asymmetric, while the displacement distribution is assumed to be Holtsmarkian. A correct method would be a convolution, but it is more practical to have some power addition rule to add two numbers that characterize the distributions. What is often used is the quadratic addition of the FWHM of the distributions (Fransen *et al.*, 1998; Van Veen *et al.*, 2001; Sakawa *et al.*, 2003), but this is a valid method only for two Gaussian distributions. According to Jansen (1990) the FW50 of two intensity distributions of different shape can be added using

$$\mathrm{FW50}_{\mathrm{total}} = \left(\mathrm{FW50}_1^{\gamma_{\mathrm{eff}}} + \mathrm{FW50}_2^{\gamma_{\mathrm{eff}}}\right)^{1/\gamma_{\mathrm{eff}}}, \tag{4.35}$$

where

$$\gamma_{\mathrm{eff}} = \frac{\gamma_1 \mathrm{FW50}_1 + \gamma_2 \mathrm{FW50}_2}{\mathrm{FW50}_1 + \mathrm{FW50}_2}. \tag{4.36}$$

γ_1 and γ_2 correspond to the shape of the distributions to be added: for example, γ_x is 2 for a Gaussian shape, 3/2 for a Holtsmark distribution, and 1 for a Lorentzian distribution. Equations (4.35) and (4.36) are used for both the virtual source size and the energy spread.

To add the FW50 trajectory displacement to the FW50 virtual source size the effective gamma is found using $\gamma_{\mathrm{blur}} = 3/2$ for the trajectory displacement (Holtsmarkian) and $\gamma_{\mathrm{intr}} = 2$ for the virtual source (Gaussian). The new brightness follows from

$$B = \frac{d_{\mathrm{intrinsic}}^2}{d_{\mathrm{blurred}}^2} B_{\mathrm{intrinsic}}. \tag{4.37}$$

To add the FW50 energy spread broadening to the FW50 intrinsic spread, $\gamma_{\mathrm{blur}} = 3/2$ for the broadening (Holtsmarkian), but what is an appropriate γ_{intr} value for the intrinsic spread? Figure 4.20 gives the shape of the energy distributions calculated for 1,800 K for a range of field strengths from 0.5 V/nm to 2.5 V/nm.

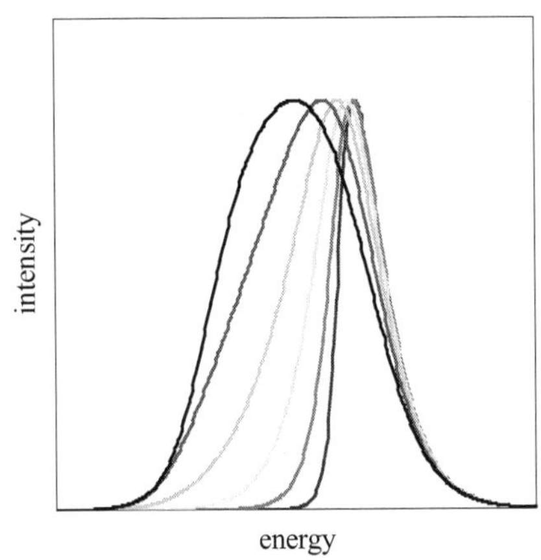

Figure 4.20 Energy distributions calculated for a surface temperature of 1,800 K and a work function of 2.93 eV and field strengths from 0.5 V/nm to 2.5 V/nm. Distributions are vertically scaled and horizontally shifted for better shape comparison.

It is found that Gaussian-shaped curves with equivalent FW50s are a reasonable fit, especially for higher field strengths. For lower field strengths a shape between Holtsmark and Gaussian would be slightly better. For simplicity it is decided to stick to $\gamma_{intr} = 2$ for all distributions. This choice will, to some extent, compensate the overestimation of the Boersch effect at low angular intensities (Fig. 4.19).

Figure 4.21 shows the dependence of the energy spread, source size, and brightness as a function of the distance from the emitting surface for the 550 nm emitter (Fig. 3.27) for different extraction voltages.

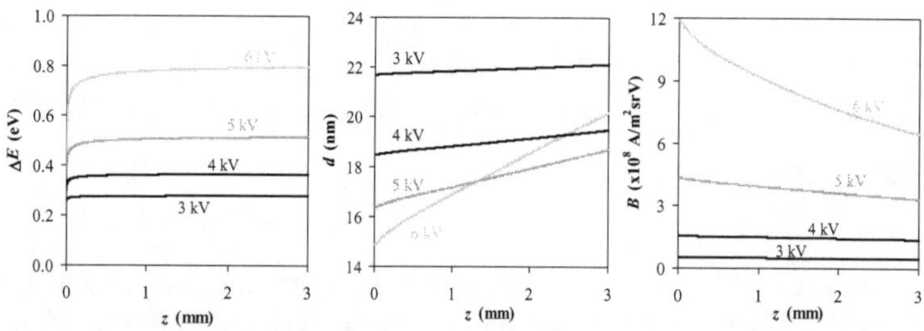

Figure 4.21 Total energy spread, virtual source size, and brightness as a function of z for the 550 nm emitter (Fig. 3.27) operating at 1,800 K and at different extraction voltages.

The initial brightness and energy spread at $z = 0$ increase with increasing extraction voltage, while the intrinsic virtual source size decreases. With respect to the effect of the Coulomb interactions, Fig. 4.18 shows that the trajectory displacement and Boersch effect increase with increasing extraction voltage. Added together, in Fig. 4.21, the net result is that the energy spread increases with increasing extraction voltage, while the brightness decreases.

An example of the effect of temperature is not given. In general it is much smaller than the effect of the extraction voltage, also due to the limited range of temperatures that can be applied in practice. Because of the temperature dependence of the work function the effect of temperature is also less straightforward than the effect of voltage.

Figure 4.22 compares the results for the different emitter geometries operating at the same temperature and with the same field strength at the facet center. The initial brightness and energy spread at $z = 0$ are thus equivalent, but the initial virtual source size is not.

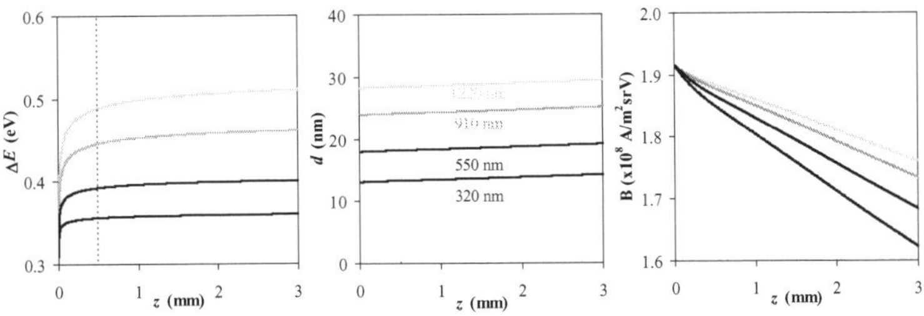

Figure 4.22 Total energy spread, virtual source size, and brightness as a function of z for the four emitter geometries of Fig. 3.27, operating at 1,800 K and with a field at the facet center of 1 V/nm. The dashed line indicates the position of the extractor.

For equivalent conditions at the center of the facet, the largest emitter suffers from the largest Boersch effect and the largest trajectory displacement (Fig. 4.19). The virtual source size is, however, also largest for the largest emitter, and the effect of the trajectory displacement is thus relatively smaller. The net result is a larger energy spread for the larger emitter but a smaller brightness loss. It is noted that the additional Boersch effect for drift spaces longer than 3 mm is very small, but the brightness will continue to decrease, for example, to 74%, 71%, 65%, and 60% at 10 mm and to 56%, 52%, 45%, and 39% at 20 mm, in order of decreasing tip size.

With respect to the practical brightness of the emitters it can be seen that the loss in the first half millimeter, between the emitter and the extractor, is small. Equation (4.17) is thus a good approximation for the practical brightness of these sources in these conditions and even more so for operation at lower field strengths. But if the current density in the beam is increased, for example, by operating at higher fields, the practical brightness of the Schottky emitter, as in the definition of Eq. (4.13), may become affected by the Coulomb interactions. It is noted that for other emitter types the effect of Coulomb interactions may actually be dominant, yielding a practical brightness largely determined by the interactions occurring between the emission site and the extractor. We expect this to be the case for the new source types based on carbon nanotubes (de Jonge, 2004) and nanotips (Vladár *et al.*, 2006).

4.4.2 The Boersch Effect Extracted from Energy Spread Data*

The energy spread of the electrons from a Schottky source that leave the extractor has been measured by a number of researchers (Kim *et al.*, 1997a; Kim *et al.*, 1997b; Kim *et al.*, 1995; Sakawa *et al.*, 2003; Bell & Swanson, 1979; Orloff *et al.*, 1985; Fransen *et al.*, 1999; Fransen *et al.*, 1998), and a contribution due to the Boersch effect generally is recognized, but up to recently only Fransen *et al.* (1998) tried to extract this contribution from their data. They estimated the Boersch effect by quadratic subtraction of the estimated FWHM of the intrinsic contribution from the FWHM of the measured energy distribution. But as mentioned, the quadratic subtraction or addition of FWHMs is a correct method only when the intrinsic distribution and the Boersch contribution both would have a Gaussian shape (Jansen, 1990), which is generally not the case. The correct way to extract the contribution from electron–electron interactions is to deconvolute the energy spread measurements with the intrinsic distribution. But deconvolution of typically noisy experimental spectra is difficult. Instead we will try to extract the Boersch effect by a method of fitting the measured spectra with convolutions of the intrinsic energy distribution and a bell-shaped distribution with free shape parameters. (After it had become available this method has been applied by Schwind *et al.* [2006] and Swanson and Schwind [2008].)

4.4.2.1 Function to represent the Boersch effect

To represent the Boersch effect we use the bell-shaped function given by Barth and Nykerk (1999):

$$v(E) = \frac{1}{(1+4(2^{1/q}-1)(E/\Delta E_{\mathrm{FWHM}})^2)^q}, \tag{4.38}$$

with electron energy E and two shape parameters, ΔE_{FWHM} and q.

To motivate this choice we compare the bell curve to the four types of displacement distributions of the extended two-particle model (Jansen, 1990; Kruit & Jansen, 2008) (Fig. 4.23).

The bell-shaped distribution of Eq. (4.38) is exactly a Lorentzian for $q = 1$, but for the other three regimes the bell is an approximation. For the best approximation we search for the q value in Eq. (4.38) that gives an FWHM/FW50 ratio of the distribution that is closest

*Most of this paragraph has been previously published and was reproduced with permission from *J. Vac. Sci. Technol. B*, **25** (2007) 2049, Copyright 2007, American Vacuum Society (see also page 249 of this book).

to that for the displacement distributions. This yields that for large values of q the bell-shaped distribution becomes approximately Gaussian, for q = 2.17 Holtsmarkian, and for q = 0.63 close to the pencil beam. The distributions of Jansen and the bell curves with the same FWHM/FW50 ratio are given in Fig. 4.23. It is noted that the bell curve with the same FWHM/FW50 ratio as a Gaussian distribution cannot be distinguished from the Gaussian curve.

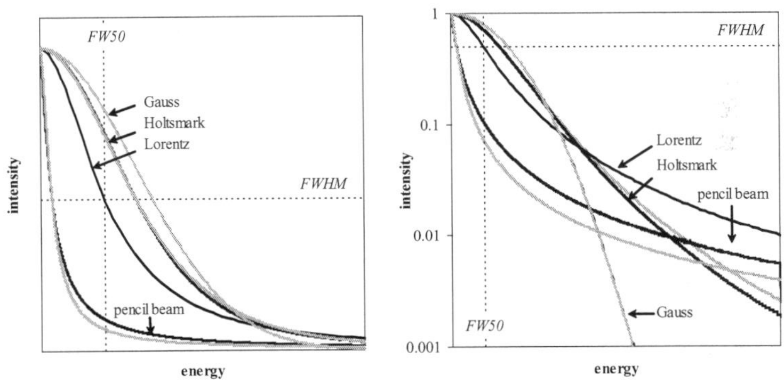

Figure 4.23 Displacement distributions of Jansen (black) and bell curves with the same FWHM/FW50 ratio (gray). Distributions have been scaled so that they have the same FW50.

Figure 4.23 shows that the bell curve is a good approximation of the Holtsmarkian and pencil beam at least down to 1% of the maximum intensity. In practice the tails of a distribution are difficult to measure accurately due to signal/noise issues. It is also noted that the assumptions in Jansen's theory may lead to an overestimation of the probability of a large displacement, meaning the tails of the resulting displacement distribution should be viewed with some caution (Jansen, 1990).

4.4.2.2 Total energy distribution measurement

The energy distributions for which the Boersch effect is extracted are energy distributions of electrons leaving a standard Schottky emitter module (Fig. 4.24). The energy distribution was measured with a retarding field energy analyzer (Simpson, 1961) in a "field-free zone" beyond the extractor. The voltage on the retarder element is varied to select electrons of specific energy. The five-element filter lens makes a crossover just after the retarder element. Additional broadening due to electron interactions in the drift space between the extractor and the analyzer and in the analyzer itself is negligibly small. The resolution of the analyzer is estimated to be better than 50 meV.

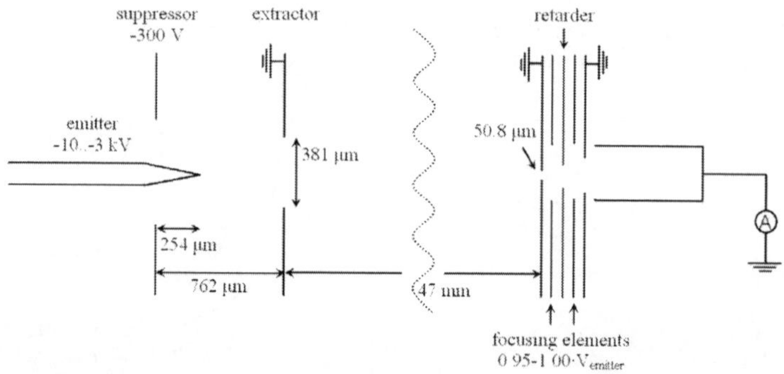

Figure 4.24 Schematic drawing of the experimental setup for measuring the total energy distribution.

With the setup in Fig. 4.24 energy distributions are recorded for four Schottky emitters of different sizes, operated at 1,800 K, and at angular intensities up to 1 mA/sr. The angular intensity was varied by varying the extraction voltage. The suppressor voltage was kept constant. Tip size is here defined as the emitter radius specified by the manufacturer.

4.4.2.3 Intrinsic contribution

The intrinsic energy distribution for each experimental distribution can be calculated when the field, temperature, and work function of the emitting surface are known—Eq. (1.13) and Section 1.4.3. The temperature of the tip was determined with a thermal imaging pyrometer (Mikron M9103 Pyrovision) using emissivity values for pure tungsten and a correction for window transmission. The field enhancement factor of the emitter and the work function have been derived from Schottky plots (Section 3.4) using Eq. (3.11) to take account of the lens effect between the emitter and the extractor. An example of a typical result is given in Fig. 4.25.

With the temperature and work function of the emitting surface and the field enhancement factor of the emitter, the intrinsic energy distribution for each experimental spectrum was generated using Eq. (1.13).

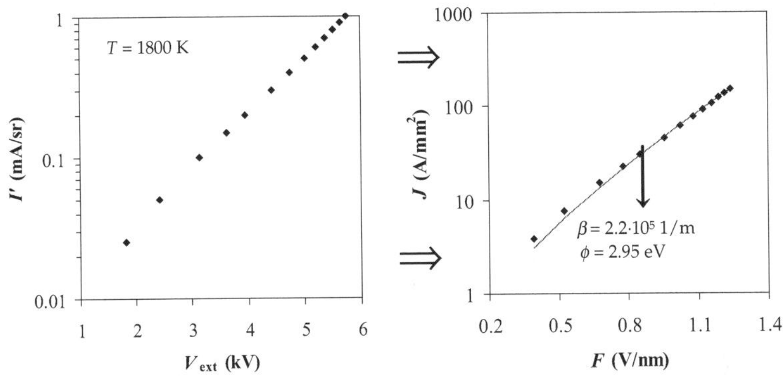

Figure 4.25 Typical example of a fit result of the theoretical $j(F)$ relation to the experimental $I'(V_{ext})$ relation used to extract the work function and field enhancement factor for the operating emitter.

4.4.2.4 Fit results

The Boersch contribution in the measured energy spectra can now by found by fitting the convolution of the intrinsic energy distribution, Eq. (1.13), and the bell of Eq. (4.38) to the experimental distribution, where the FWHM and $1/q$ shape parameters of the bell are used for fitting (step size FWHM 0.01 eV, $1/q$ 0.05). The maximum of each experimental spectrum is fitted with a parabola through its top to align it with the convolution and is allowed to shift up to 2.5% vertically (but not above the maximum experimental point) and 0.025 eV horizontally. To find a convolution point for every experimental point linear interpolation is used. The fit criterion is the least squares.

Because the main interest is the effect of the choices in operating conditions and tip geometry on the contribution from chromatic aberration, the fitted FWHM and $1/q$ values of the bell curve are converted into an FW50 value. The FW50 value is directly proportional to the chromatic aberration, independent of the shape of the distribution, unlike the FWHM (Barth & Nykerk, 1999). For a Schottky emitter in normal operation it turns out that the FW50 of an energy distribution is ~0.6–0.7 × FWHM, depending on the conditions.

Typical fit results for the experimental energy distributions are given in Fig. 4.26. It demonstrates the quality of the fit and shows that the Boersch effect increases for increasing extraction voltage (angular intensity). The effect of increasing extraction voltage (increasing intrinsic energy spread) on the Boersch effect contribution is made more quantitative in Fig. 4.27.

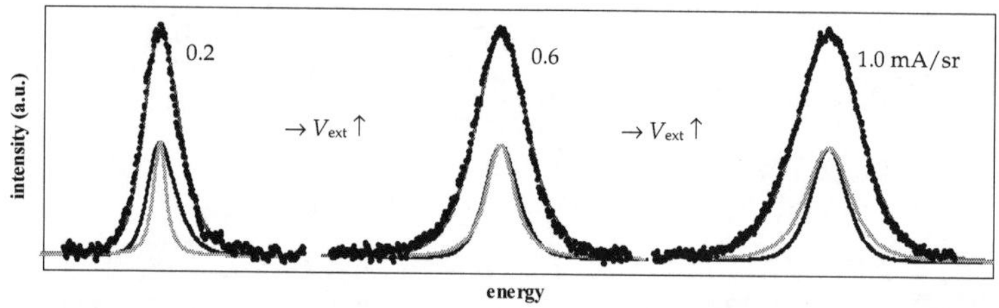

Figure 4.26 Typical fit results obtained by fitting the convolution of the intrinsic energy distribution and a bell-shaped function with two fit parameters to the experimental energy distribution (bullets: data; solid line: fitted convolution). Also given are the intrinsic (black) and Boersch effect distributions (gray) scaled to half the maximum of their convolution. Data are taken for a Schottky electron source with a tip radius of 0.55 μm operated in the experimental setup of Fig. 4.24 at 1,800 K at different extraction voltages to give an angular intensity of 0.2 mA/sr, 0.6 mA/sr, and 1.0 mA/sr.

While the Boersch effect contribution increases with increasing intrinsic energy spread, the shape of the bell that represents the Boersch effect changes from more Lorentzian toward more Holtsmarkian, in agreement with the predictions of Fransen *et al.* (1999). Because the noise in the spectrum increases with decreasing extraction voltage, the uncertainty of the $1/q$ fit result is largest for the lowest intrinsic energy spread. Assuming our intrinsic distribution is correct, Fig. 4.27 also shows the resolution of the analyzer is high indeed because at low current densities, the broadening (which includes the analyzer resolution) goes below 50 meV.

Results like those of Fig. 4.27 were obtained for tips of different sizes. In published energy spread data, both experimental and calculated, the results for different tip radii are often presented in graphs of the FWHM of the total energy distribution versus angular intensity (Kim *et al.*, 1997b; Kim *et al.*, 1995; Sakawa *et al.*, 2003; Bell & Swanson, 1979; Orloff *et al.*, 1985; Thomson, 1994). This is probably because the FWHM of a distribution is easy to determine and the angular intensity is relatively easy to measure. Such plots are useful when one is interested in which emitter gives the lowest total energy spread at a particular angular intensity. An example has been given in Fig. 4.28a (data from Schwind *et al.* [2006]), and it can be seen that a larger emitter has a lower total energy spread at equivalent angular intensity.

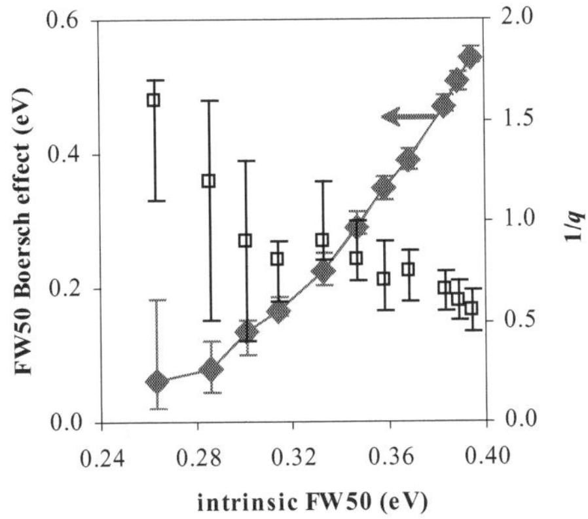

Figure 4.27 Fit result for the FW50 of the Boersch effect and the $1/q$ value versus the intrinsic energy spread for a Schottky electron source with a tip radius of 0.55 μm, operated in the standard gun geometry of Fig. 4.24 at 1,800 K and at different extraction voltages, to give an angular intensity of 0.05–1.00 mA/sr. The error bar for $1/q$ indicates the $1/q$ values that (in combination with an FWHM value) gave a fit result up to 10% less good than the best fit. The error bar for the FW50 indicates the FW50 values associated with these combinations of $1/q$ and FWHM.

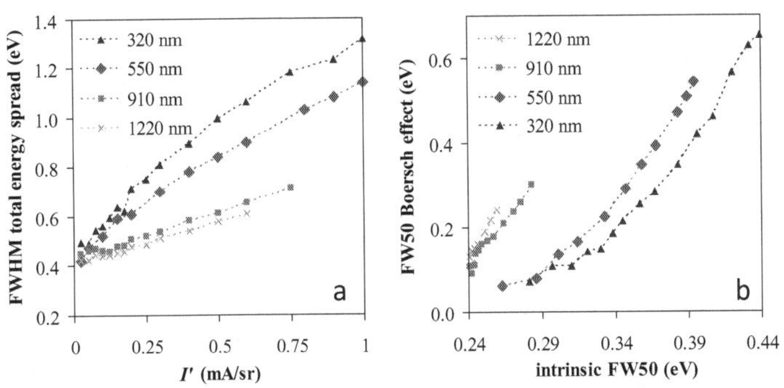

Figure 4.28 (a) FWHM of the total energy distribution versus angular intensity for different emitter sizes (radius in nm) (Schwind *et al.*, 2006). (b) Extracted FW50 of the Boersch effect versus intrinsic FW50 energy spread for the same data set. Schottky emitters at 1,800 K in the experimental setup of Fig. 4.24.

This may suggest the larger emitter suffers less from the Boersch effect, while the opposite was concluded from the simulations in the previous section. This "contradiction" is only apparent:

Figure 4.28b is based on the same data set as Fig. 4.28a but gives the extracted Boersch effect contribution as a function of intrinsic energy spread. It shows that for equivalent intrinsic energy spread a larger emitter has a larger Boersch effect contribution in agreement with the simulation results. The reason why in Fig. 4.28a, at equivalent angular intensity, the total energy spread of a larger emitter is smaller than that of a smaller emitter is because for the larger emitter the intrinsic energy spread component in the total energy spread is much smaller than for the smaller emitter. This is a consequence of the difference in the emitter–extractor lens effect for different emitter sizes (Fig. 3.30): for larger emitters the lens effect is less diverging. The field strength must thus be much lower for a larger emitter to have the same angular intensity as a smaller emitter. Comparing at equivalent field strength, yielding an equivalent intrinsic energy spread (Fig. 4.28b) and equivalent practical brightness, the larger emitter will have a larger total energy spread. Because the practical brightness, and not the angular intensity of emitters (but their angular intensity divided by the square of their virtual source size and the extraction voltage—Eq. (4.13)), is directly proportional to the source image size or spatial beam coherence, it is more useful from an application point of view to compare the energy spread of emitters (affecting the chromatic aberration) for equivalent practical brightness.

4.4.2.5 Comparison with theory

We compare the results of the deconvolution approach with the results from Eqs. (4.32)–(4.34). For the half opening angle α_{ext} we take 7º again, and we take $0.3 \times r_{tip}$ as the crossover radius r_c. Figure 4.29 shows that the result from Eqs. (4.32)–(4.34) are in reasonable agreement with the experimental results. While the larger emitters are operating in the Holtsmark regime, the smaller emitters show a transition toward the Lorentz regime for decreasing angular intensity. It is noted that applying Knauer's formula (Knauer, 1981) (an approximation that assumes a spherical electric field around the emitter; see for a comparison, e.g., Jansen [1990] or Fransen *et al.* [1999]) yields values exceeding those of Jansen by a factor of 1.5–2.7 (equations for the FWHM of the energy spread).

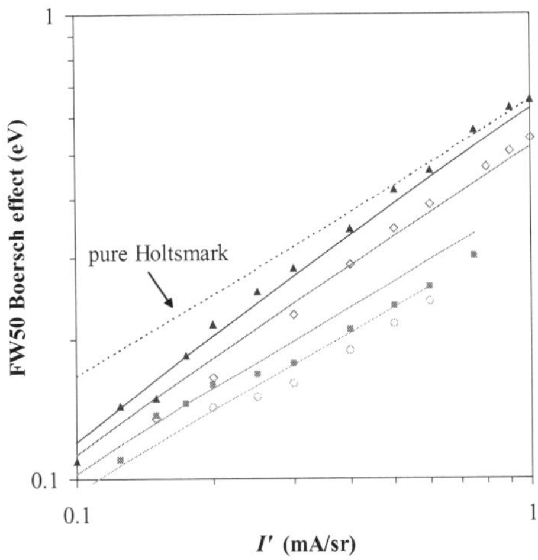

Figure 4.29 Comparison of the FW50 of the Boersch effect extracted from experimental spectra (markers) and calculated with Eqs. (4.32)–(4.34). Four Schottky emitters of different radii (nm) at 1,800 K in the experimental setup of Fig. 4.24.

4.4.2.6 Discussion

Important for a good fit result are experimental spectra with a sufficiently large range and high resolution and an accurate intrinsic energy distribution. If too small a part of the tail of the experimental distribution has been measured the fit becomes insensitive to the shape of the Boersch effect distribution, although not to the FW50 of the Boersch effect. This shows up in Fig. 4.27, where the error in $1/q$ is large, especially at lower intrinsic energy spread, while the error in the FW50 value remains small.

For an accurate intrinsic energy distribution an accurate temperature, work function, and field are required. To make sure the emitter shape, and thus the field enhancement factor, would be the same for all measurements, the I'–V_{ext} data and the energy distribution have been taken shortly after each other.

It is noted that the I'–V_{ext} data are taken for a constant suppressor voltage and thus for a variable field form. This will introduce an error in the intrinsic energy spread—especially for low extraction voltages the spread will be overestimated. As a consequence the Boersch effect at the lowest extraction voltages (angular intensities) will be underestimated. The effect is, however, relatively small:

taking a 10% lower field enhancement factor changes the best fit FW50 Boersch effect for the energy distribution from a 550 nm radius Schottky source at 1,800 K and 0.2 mA/sr from 0.17 eV to 0.19 eV.

The method was demonstrated for energy distributions of electrons emitted by Schottky sources but can be used for spectra recorded from other emitter types as well. Possibly the method could also be used to extract the trajectory displacement contribution from probe intensity profiles, which could be of use in the design of guns for higher practical brightness.

4.5 Summarizing: The Beam Properties Relevant to Electron Optical Systems

On the basis of the preceding paragraphs we can now quantify the brightness, energy spread, and virtual source size of a beam from a Schottky emitter coming through the extractor aperture and fully characterize the performance of focused electron beam systems.

Figure 4.30 gives for the four emitters in Fig. 3.27, in the standard gun geometry, with the suppressor voltage at a constant –300 V, the relation between energy spread and brightness at three different temperatures at a distance of 1 mm from the emitting surface. This position could, for example, represent the position of the beam-defining aperture that minimizes further brightness loss and energy spread broadening in the column.

Figure 4.30 shows that there is theoretical maximum in the brightness, as was also shown for 1,800 K by Van Veen *et al.* (2001), based on a slightly different calculation. The added value of this graph is that it includes an estimate of the corresponding energy spread, as well as the effect of temperature. The maximum brightness is the result of the effect of Coulomb interactions, which increases with increasing intrinsic brightness. The theoretical brightness maximum of larger emitters is larger but at the same time associated with a larger energy spread. The figure has two sets of five markers. Each set has the same intrinsic energy spread and brightness. The markers show that at lower intrinsic brightness there is energy spread broadening but no significant brightness loss, while at higher intrinsic brightness both energy spread broadening and brightness loss take place. They also make clear that the energy spread broadening is strongest for the largest emitter, while the brightness loss is the smallest. Which emitter at which conditions will give the best performance depends on the specific application. For example, sometimes high brightness is more important and sometimes a low energy spread. This will be addressed in more detail in Chapter 6.

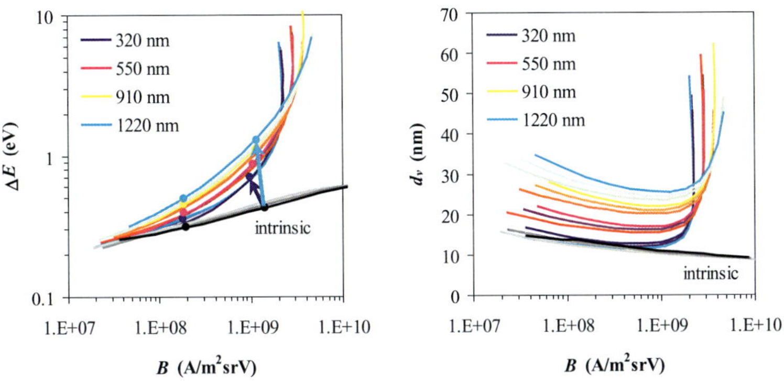

Figure 4.30 Relation between energy spread (FW50), practical brightness, and FW50 virtual source size at 1 mm from the emitting surface for four different emitters in the standard gun geometry at three different temperatures (1,600 K, 1,700 K, 1,800 K). The work function at 1,800 K is 2.93 eV. In the right graph only the intrinsic $B(d_v)$ relation for the 320 nm emitter is given.

For multibeam applications it is of interest to know how the brightness, energy spread, and virtual source size vary for different selections of the facet beam. This is shown in Fig. 4.31 for different voltage settings, for the emitter geometry in Fig. 3.11, in the standard gun geometry, with a suppressor–extractor distance of 0.76 mm, a protrusion of 0.28 mm, and a fixed suppressor voltage of –0.3 kV with respect to the emitter, which is at 1,800 K and has a work function of 2.93 eV.

Note that Fig. 4.31 shows intrinsic properties without the effect of Coulomb interactions.

Figure 4.31 shows that an off-axis beamlet will have a higher brightness and energy spread and a smaller virtual source size than the central beamlet. And the angular intensity of the off-axis beamlet can be both higher and lower than that of the central beamlet, depending on the voltage settings.

Due to the nonuniformity of the field across the facet, and due to the aberrations of the lens between the facet and the extractor, it is likely that the effect of Coulomb interactions is different for off-axis beamlets than for the center of the facet beam. A quantification of the effect of Coulomb interactions for off-axis beamlets would require a full-particle analysis because approximations of constant particle density within the relevant interaction range are no longer valid. But in general, due to the lens aberrations, the beam is spread out more toward its edges, and this is likely to give a relatively smaller contribution from Coulomb interactions as compared to a more compressed beam such as near the beam center.

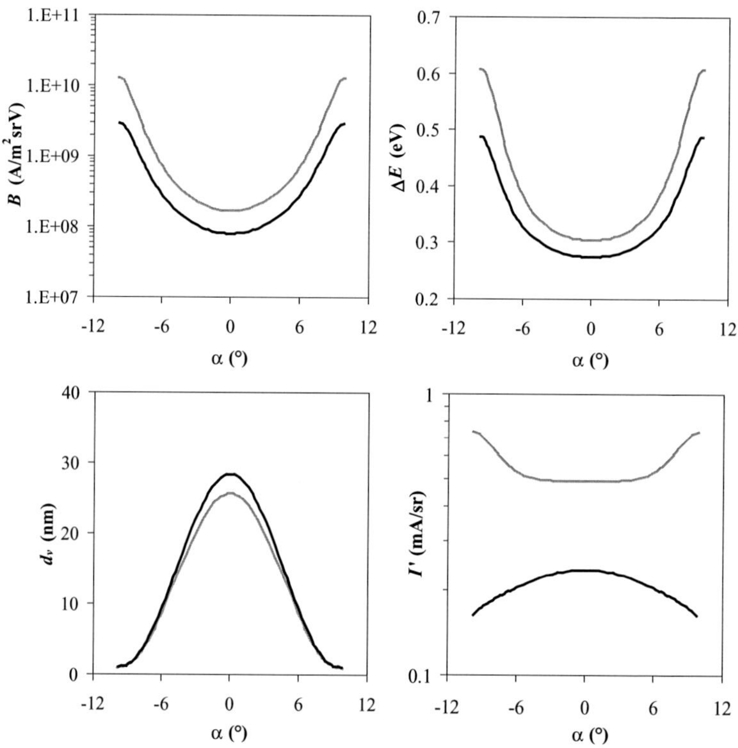

Figure 4.31 Nonuniformity in intrinsic brightness, energy spread, virtual source size (first approximation for beamlets with very small opening angles), and angular intensity distribution for an extraction voltage of 4.75 kV (black) and 5.75 kV.

The nonuniformity of the properties across the facet beam poses both opportunities and challenges. It offers an opportunity to change the beam properties at the target without changing the operating conditions of the electron source, simply by switching between the center and an outer beamlet. (Note that this does require adaption of the standard electron optics, e.g., deflecting the facet beam to select a different part with the same aperture or adding an off-axis aperture and the necessary optics to switch between beamlets) (Young *et al.*, 2009). In this way one could switch between different applications (e.g., high-resolution imaging and chemical analysis). Note that although the brightness and energy spread are larger for an off-axis beamlet, the current density toward the outer perimeter of the beam is strongly dependent on the operating conditions and can be both higher and lower than the current density in the beam center. This is relevant with respect to the desired (change in) beam current and the acceptance angles of the aperture(s).

The inherent nonuniformity is more of a challenge for the design of multibeam systems in which the facet beam is split into multiple beamlets (with, e.g., an aperture plate), which preferably all have equivalent properties (Zhang & Kruit, 2007).

Relevant to both multibeam and single-beam systems is the stability of the physical shape of the end facet. This will be addressed in the next chapter.

Chapter 5

Geometrical Stability

Emitter shape is an important parameter for performance of the source, because it affects the field strength at its surface and the lens effect between the emitter and the extractor. The shape is, however, generally not in thermodynamic equilibrium, and as a consequence the source can evolve and display a performance that is not constant in time.

Shape changes become particularly important when probe properties need to be constant over a long time period (lithography applications) and/or when a larger area of the end facet is of interest, such as for multibeam applications (Zhang & Kruit, 2007; Young *et al.*, 2009).

Below we will address different consequences of thermodynamic instability: collapsing of the end facet, and other geometry changes of the end facet, and changes of the emitter on a larger scale—the overall size of the tip—and how this affects source performance.

5.1 Observed Geometrical Changes

The collapsing of the end facet is historically known as "collapsing rings" (Trolan *et al.*, 1955). The phenomenon of collapsing rings as a shape change mechanism for electron sources was first reported by Trolan *et al.* (1955) for hot single-crystalline tungsten field emitters. The terminology originates from the evolution of the electron emission patterns showing contracting bright rings. The bright rings

Physics of Schottky Electron Sources: Theory and Optimum Operation
Merijn Bronsgeest

ISBN 978-981-4364-79-9 (Hardcover), 978-981-4364-80-5 (eBook)
www.panstanford.com

were associated with enhanced emission from the sharp edge of circular {110} islands of an atomic height centered on the tip end. The contraction was associated with a size reduction of the island caused by net detachment of atoms from the island. The detached atoms diffuse away and reattach elsewhere on the emitter. Figure 5.1 gives an example of rings in the emission pattern of a collapsing Schottky emitter. Inspection of Schottky emitters frozen in the middle of such an instability shows islands on the facet with a height of up to tens of nanometers (Fig. 5.2) and thus not the atomic steps as seen for tungsten {110} emitters.

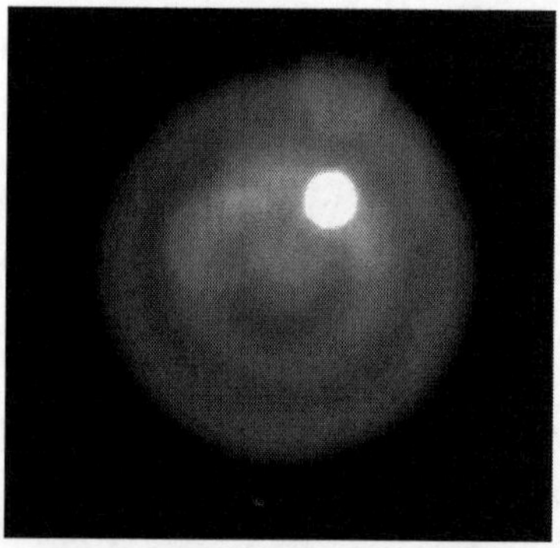

Figure 5.1 Example of bright rings in the emission pattern of a Schottky emitter. The bright spot is not caused by electrons, but it is light coming from the hot source, penetrating the semitransparent YAG screen. *Abbreviation*: YAG, yttrium-aluminum-garnet.

The collapsing emitters were found to become thicker and shorter over time. This was explained theoretically shortly after discovering the collapses. This theory is addressed in Section 5.3.1 and predicts the evolution of the emitter macroscopically for isotropic material properties. A model that can actually describe the phenomenon of collapsing rings on a more basic level, microscopically, will be discussed later in Section 5.5.1.

It has been shown that collapsing can be stopped by applying a high-enough field. This is addressed in both models (Sections 5.3.1 and 5.5.1). The fact that collapsing can be prevented by applying a field may suggest also that tip "thickening" can be prevented. This seems, however, not to be the case. Figure 5.3 shows emitters as produced and after operating them in different conditions. Tip size growth without the collapses is addressed in Section 5.3.2.

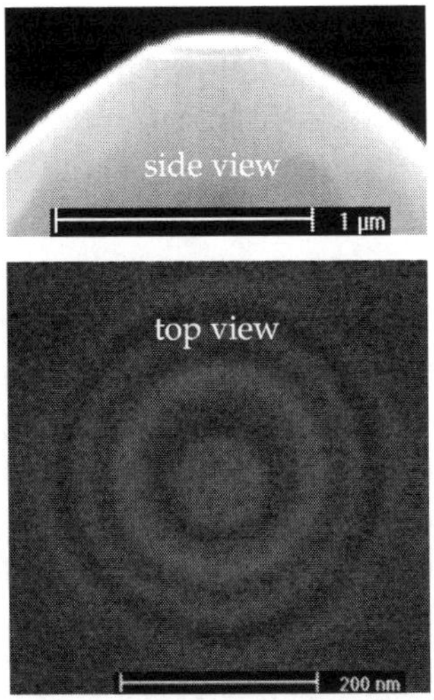

Figure 5.2 Examples of SEM images of Schottky emitters frozen in the collapsing process. (The emitters' geometry is "frozen" by switching off the heating current. This causes the emitters to cool down, through radiation and conduction, almost instantly to a temperature at which surface diffusion processes are already slow enough to prevent changes on the scale of the features shown.) *Abbreviation*: SEM, scanning electron microscopy.

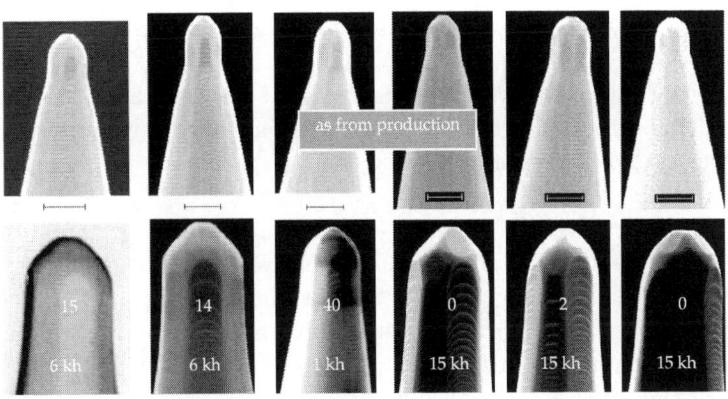

Figure 5.3 Schottky emitter tips as produced and after a number of operating hours (in kh). Also indicated is the number of collapse events for the end facet. The scale bars are 1 µm. (Most images from FEI Company.) *Abbreviation*: kh, kilohour.

The third observation concerns the shape of the end facet. The top view of the end facet in Fig. 5.2 is circular, but this is not necessarily the case, as could already be concluded from the experimental emission patterns in Chapter 3. In confirmation, Fig. 5.4 gives a collection of SEM images showing different geometries.

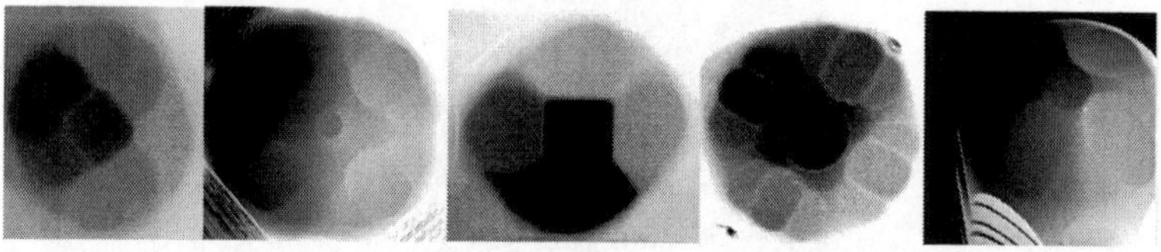

Figure 5.4 Top-down SEM images of Schottky emitters with different facet geometries, generated by operating at different conditions.

The effect of operating conditions on the end facet geometry is discussed in Sections 5.2 and 5.4.

The effect of geometric changes on the properties of the beam for applications are most obvious for the phenomenon of the collapsing end facet. Collapsing rings cause serious probe instabilities, such as in Fig. 5.5. The beam current can drop as much as ~70%. These fluctuations do not occur if the field strength at the emitter is high enough, but a high electric field is also associated with a relatively high energy spread of the electron beam. For applications that are limited by the energy spread, the source is usually operated at extraction voltages that give reduced fields, and beam instabilities, as in Fig. 5.5, do become a risk.

The origin of the typical profile of Fig. 5.5 has never been clarified in detail. The details of collapsing rings on a Schottky emitter are investigated in Section 5.5.4, and their correlation with the probe are investigated in Section 5.6.2.

The effect of changing facet geometry on the source performance is less obvious than of the collapse of the facet. But if the facet geometry is changing in time, the brightness and energy spread of the beam are a function of time, too. This should be taken into account if constant beam properties are crucial, as may be for, for example, lithographic applications. This will be discussed in Section 5.6.1.

The overall growth of the tip that is observed, irrespective of the conditions, reduces the field enhancement at the tip. This means that without changing the extraction voltage, the intrinsic brightness and energy spread will decrease during the lifetime of the emitter.

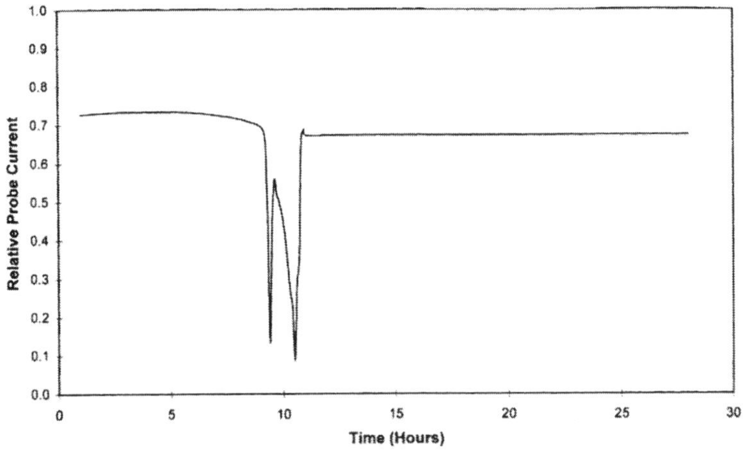

Figure 5.5 Typical periodic beam instability associated with collapsing rings (from Swanson and Schwind [2008]).

5.2 Equilibrium Crystal Shapes

To explain the geometric changes that were introduced above, it is convenient to treat the emitter as a crystalline, pointed shape that is supporting a "crystallite," being the tip, and to distinguish "global" atom transport, between the tip area and the cone, and local rearrangement in the tip area itself.

In general, a piece of crystalline material will want to evolve to its equilibrium crystal shape, which is the shape that gives the lowest total surface free energy for the given circumstances. The total surface free energy is the surface integral of the specific surface free energy (surface free energy per unit area) and a function of the "natural" variables (those that are kept constant) of the thermodynamic system considered, which can be, for example, temperature, volume, and applied potential.

If the specific surface free energy is a constant, and independent of surface orientation, the equilibrium crystal shape of a given volume of material is a sphere: a sphere is the solution with the smallest surface area. But the specific surface free energy (the surface free energy per unit area) is generally different for different crystallographic orientations. To first approximation this can be understood from the broken bond model that was introduced by Mackenzie (1962) but is still used today (Liu *et al.*, 2009): the surface energy of a crystallographic orientation can be found by calculating the energy involved in breaking the bonds (between nearest neighbors, between second-nearest neighbors, and so on) upon cleavage.

Surface energy, or surface tension, is difficult to determine, and in the literature there exists a considerable scatter in data, both from calculations (e.g., Skriver & Rosengaard, 1992; Chen, 1992; Rodriguez *et al.*, 1993; Vitos *et al.*, 1998; Hu *et al.*, 2002; Zhang *et al.*, 2003; Wen & Zhang, 2008) and from experiments (e.g., Barbour *et al.*, 1960), and an overview in Rodriguez *et al.* (1993). But when the dependence of the specific surface free energy on the crystallographic orientation is known, the equilibrium shape of a crystallite can be found from a so-called Wulff plot or γ-plot. The Wulff construction (Fig. 5.6) of the equilibrium shape is obtained by drawing a plane through each point on the γ-plot perpendicular to the line connecting that point to the origin. The body formed by all points reachable from the origin without crossing any of these planes is geometrically similar to the equilibrium shape.

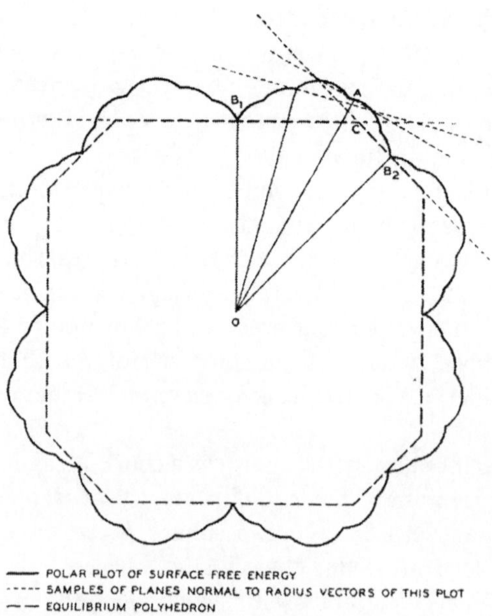

Figure 5.6 Example of the Wulff construction in 2D (taken from Herring [1951]). *Abbreviation*: 2D, two-dimensional.

The Wulff plot depends on the applied conditions. Upon heating a crystallite the anisotropy in specific surface energy will diminish: The specific surface free energy reduces for an increasing entropy contribution, and with increasing temperature the (configurational) entropy increases more for less densely packed planes (Williams, 1992). As a result the equilibrium shape will become more spherical (e.g., Ibach, 2006). Upon applying a potential the anisotropy in the specific surface energy will increase because the specific surface free energy, independent of orientation, decreases with an amount proportional to the energy density of the electric field (Fujita & Shimoyama, 2007; Lozovoi & Alavi, 2003), and as a result the equilibrium shape will become less spherical (more faceted). The specific surface free energy also depends on the composition and coverage of adsorbed species. The Wulff plot for Schottky emitter material is unknown and might be different from the Wulff plot of bare tungsten due to the presence of zirconium and oxygen on the surface. Figure 5.7 gives a (part of 2D) Wulff plot using calculated data for bare tungsten to illustrate the effect of the temperature and field.

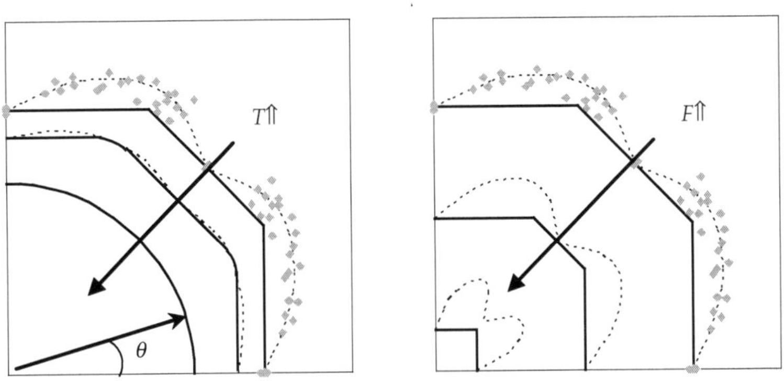

Figure 5.7 The effect of the temperature and applied field on the equilibrium shape (Wulff plot). The markers are calculated data for the specific surface free energy of tungsten (Wen & Zhang, 2008). The angle θ is with respect to the (110) plane.

Figure 5.7 shows that upon increasing temperature the equilibrium shape changes from polyhedral to a sphere, while upon increasing the field the equilibrium shape changes from polyhedral to a cube. Figure 5.8 gives an example of the effect of applying a field to a polyhedron with rounded corners. It can be seen that upon applying a field the rounded sections on the equilibrium shape can disappear, similar as for cooling down the crystal.

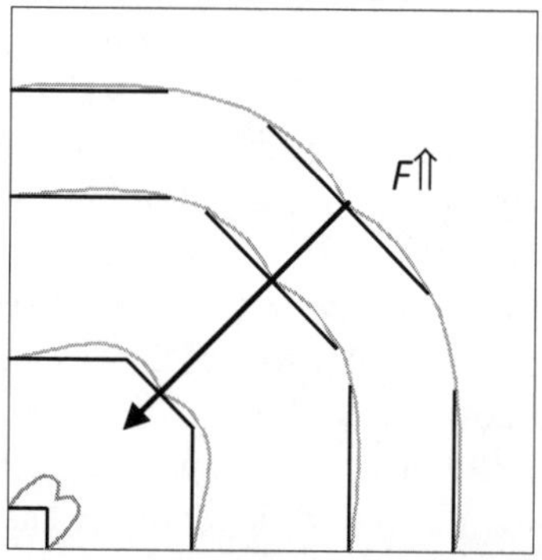

Figure 5.8 Effect of applying a field on the equilibrium crystal shape for the crystal of Fig. 5.7 at elevated temperatures.

Experimental evidence of the effect of temperature is, for example, the study on a Pb crystal (Bombis *et al.*, 2002), and for the effect of the field strength, the study of the tip end of a W{100} emitter (Fujita & Shimoyama, 2007).

To explain changes of the geometry it is sometimes enough to consider only the equilibrium shape of the tip itself, and sometimes it can be enough to consider only the equilibrium shape of the emitter as a whole. For example, the phenomenon of recession and tip size growth of pointed shapes at elevated temperature can be understood as a method to evolve toward a shape with less surface area. And the different tip-end geometries in Fig. 5.4, for different operating conditions, can be directly coupled to changes in the equilibrium shape of the tip itself.

Figure 5.9 gives a series of top-down SEM images of Schottky emitter tips taken after prolonged operation and sorted in order of increasing operating field. This order was determined by analyzing the operating conditions, and the measured angular intensity, before the emitter was cooled down, in combination with the approximate tip size. Detailed charge density method simulations have not been performed, and the ordering is thus an approximation. The general trend is, however, clear and in agreement with what can be expected on the basis of Fig. 5.8: for higher fields the tip is more faceted, while for lower fields the shape is more rounded (and the facets smaller).

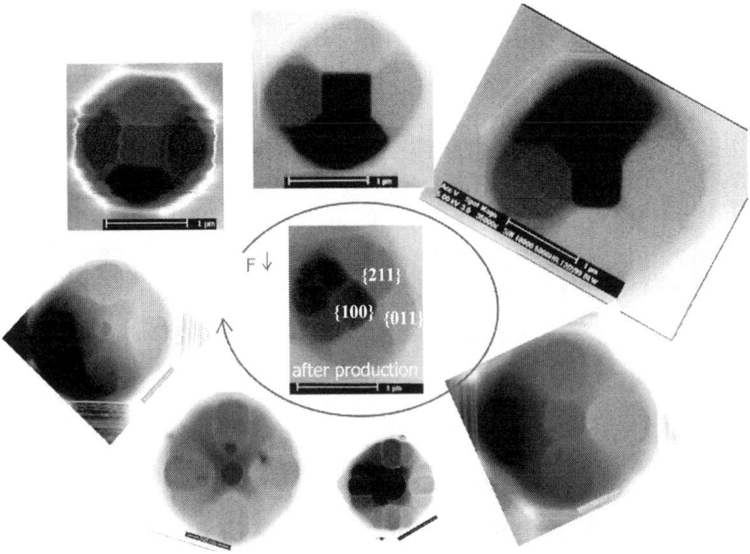

Figure 5.9 Top-down SEM images of different Schottky emitters operated at different field strengths F. The arrow indicates the direction of decreasing field strength.

It is noted that the equilibrium crystal shape of a volume is often not achieved. This has been associated with energy barriers that have to be overcome to get from the given state to the equilibrium state (Degawa *et al.*, 2006a; Mullins & Rohrer, 2000), and it can be shown that the height of these barriers increases with increasing crystal volume. This also applies to an emitter tip, but an emitter tip is a "supported" and not a "free" crystallite: there is net atom transport possible between the crystallite and the "support structure"—the emitter cone. It is noted that already in 1965 Nichols and Mullins (1965) report of a discussion in the literature as to how closely the shape near the end of an emitter tip approximates the true equilibrium shape that would be assumed by an isolated particle of the emitter material. As will be shown, the interplay between tip thermodynamics and that of the emitter as a whole is an important factor.

In extreme cases kinetics can actually start dictating the shape of a piece of material: it can be seen that in Fig. 5.7, for very high temperatures and fields, the surface free energy will go to zero (Fujita & Shimoyama, 2007; Pavlov, 2006). In such conditions the crystal equilibrium shape is no longer defined, and if the material does not evaporate, the actual shape will depend on the kinetics. It has been shown that for very high (positive and negative) fields microprotrusions can form on the low index faces of emitter tips (Shrednik *et al.*, 1974; Pavlov, 2006; Fujita & Shimoyama, 2007). To date, the mechanism of this protrusion formation has been only partly established (Fursey, 2005; Pavlov, 2006). Figure 5.10 is believed to be an example of such microprotrusions formed on the tip of a Schottky emitter, with outgrowths on the low index {110} faces on the tip end and on the neck.

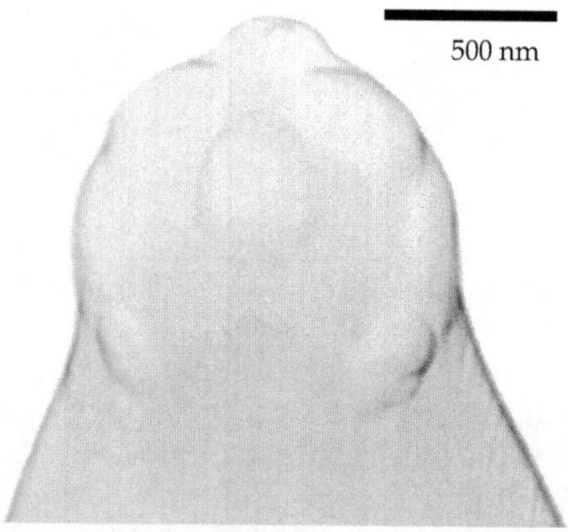

Figure 5.10 SEM image of outgrowths on the tip of a Schottky emitter suspected to be produced by operation in very high fields. (Image from FEI Company.)

5.3 Tip Size Growth

5.3.1 The Continuum Model: Tip Size Growth at Low Fields

Soon after collapsing rings on emitters were first discovered (Trolan *et al.*, 1955; Barbour *et al.*, 1960) the effect of it—the blunting and recession of the tip—was explained by considering the thermodynamics of the emitter as a whole. This approach still serves as the theoretical ground for a practical equation that predicts if a

Schottky emitter is likely to collapse or not (Swanson & Schwind, 2008).

The basic requirement in that theory is that bulk and surface are in local thermodynamic equilibrium, so a local chemical potential can be assigned to the surface. The surface chemical potential was described as an analytical function of the local surface curvature. Following Barbour *et al.* (1960) and Nichols and Mullins (1965), the surface chemical potential may be written as $\mu = \Omega_0 \gamma \kappa$, with Ω_0 the atomic volume, γ the isotropic (average) surface tension, and κ the local curvature. Along a surface with curvature variations, the associated variations in chemical potential will drive mass transport. For an emitter the direction of net diffusion is pointing away from the apex, because the curvature along the surface decreases going from the apex toward the shanks (left image in Fig. 5.11), and as a result the emitter will recede. The atoms of collapsing layers are redistributed and add to the diameter of the emitter tip.

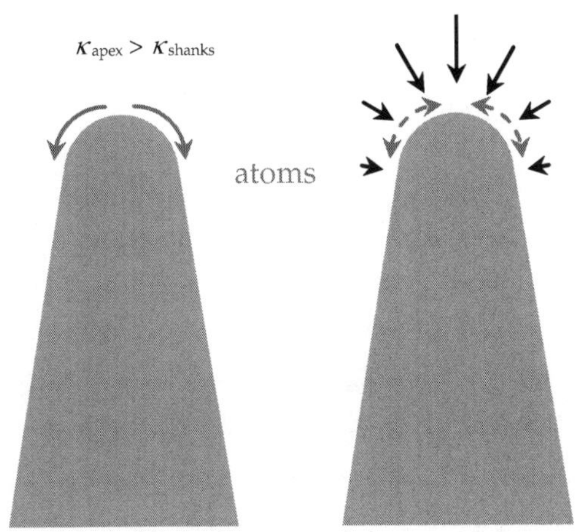

Figure 5.11 Illustration of the direction of net atom transport in the tip area in the case of a zero field (left) and in the case of an electric field equal to the "stopping field"—Eq. (5.6).

Assuming mass transport is through surface diffusion only, the local atom flux can be found from

$$J = -\frac{D_s}{A_0 k_B T}\nabla\mu = -\frac{\Omega_0^2 D_s}{A_0 k_B T}\gamma\nabla\kappa, \tag{5.1}$$

with D_s is the diffusion constant and A_0 the atomic area. Using a finite-difference method Nichols and Mullins (1965) have calculated the evolution of conical shapes (emitters) and cylindrical rods for isotropic diffusion parameters and surface tension. For rods and emitters with cone half-angles < 3º they predicted spheroidization, but for cone half-angles larger than ~3º (standard Schottky emitter typically ~14º), they calculated that a steady-state profile will be adopted from whereon the emitter will recede and the tip will increase in diameter but the emitter as a whole will maintain the same profile. It is noted that they calculated the formation of a neck behind the tip for cone angles smaller than ~8º. Necks are observed for Schottky emitters of small emitter radii but are generally not present for large emitter radii. It is suggested that the neck is a remnant of the production process, which disappears for prolonged operation of the emitter.

Once the steady-state profile is adopted, the tip radius and position can be found from Herring's scaling laws (Herring, 1951)

$$r_{\mathrm{tip},2}^4 - r_{\mathrm{tip},1}^4 = \mathrm{A}_\theta \frac{\Omega_0^2 D_s \gamma}{A_0 \mathrm{k_B} T}(t_2 - t_1), \tag{5.2}$$

$$z_2^4 - z_1^4 = \mathrm{B}_\theta \frac{\Omega_0^2 D_s \gamma}{A_0 \mathrm{k_B} T}(t_2 - t_1), \tag{5.3}$$

where A_θ and B_θ are geometry-dependent constants and $t_2 - t_1$ indicates the time step. z refers to the point of projection of the conical shank onto the cone axis. This equation can be rewritten to give the recession rate

$$\frac{\mathrm{d}z}{\mathrm{d}t} = \mathrm{C}_\theta \frac{\Omega_0^2 D_s \gamma}{A_0 k_B T} \frac{1}{r_{\mathrm{tip}}^3}, \tag{5.4}$$

with C_θ a geometry-dependent constant. This expression has been exploited experimentally to find diffusion parameters for a number of different materials (Barbour *et al.*, 1960; Bettler *et al.*, 1974). The recession rate is relatively easy to measure (by counting the number of rings collapsing per unit time in the emission pattern generated by operating a heated emitter in pulsed field mode [Barbour *et al.*, 1960] and assuming each ring is associated with a single atomic layer) and was often found to be approximately constant. The latter is because for the close to micron-size tips the increase in tip radius associated with the length reduction per collapse is approximately negligible.

What is the effect of the operating conditions? A change in emitter temperature will change the rate of diffusion processes but will not affect the tip–shank gradient in chemical potential, which is dictated by the curvature difference between the tip and the base. (It can, however, cause changes in the tip area itself because temperature affects the equilibrium crystal shape [Fig. 5.7].) Applying an extraction voltage does change the tip–shank gradient (Barbour *et al.*, 1960). By applying a voltage to the emitter the surface chemical potential is no longer a function of the local slope alone but also of the energy density of the local electric field: the free energy change per atom upon locally adding a small volume is reduced with the electric field energy stored in the vacuum volume it replaces:

$$\mu = \Omega\left(\gamma\kappa - \frac{1}{2}\varepsilon_0 F^2\right) \tag{5.5}$$

The bias between an emitter and an extractor creates a field gradient along the surface of the emitter, with the highest field at the tip, where the curvature is strongest (Fig. 2.2). From Eq. (5.5) it follows that when the relative variation in curvature in a given area is equal to the relative variation in electric energy density, there is no local gradient in chemical potential and thus no net mass transport. Barbour *et al.* (1960) analyzed the apex area of tungsten emitters and concluded from field calculations for sphere-on-cone geometries, and inspection of emitter profiles on electron microscope shadowgraphs, that there is no chemical potential gradient in the apex area for a field strength at the apex of

$$F_{\text{stop}} = 2\sqrt{\frac{\gamma/2}{\varepsilon_0 r_{\text{tip}}}}. \tag{5.6}$$

With Eq. (5.6), the recession rate is predicted to decrease for increasing field strength up to $F = F_{\text{stop}}$:

$$\frac{dz}{dt} = \left(\frac{dz}{dt}\right)_{E=0}\left(1 - \frac{1}{2}\varepsilon_0 F^2 \frac{r_{\text{tip}}}{\gamma}\right) \tag{5.7}$$

The reduction in recession rate with increasing applied voltage was confirmed experimentally and has been exploited to determine the (average) surface tension for emitters with collapsing planes of W(110) and Re(10-10) (Barbour *et al.*, 1960; Bettler & Barnes, 1968) and for the Schottky emitter (Swanson & Schwind, 2008). It is noted that for emitters with collapsing planes of Rh(111), Ir(111), and Re(0001) Eq. (5.7) did not apply at all (Bettler & Barnes, 1968).

Swanson and Schwind (2008) have experimentally determined a stopping field of 0.8 V/nm for a Schottky emitter with a reported radius of 0.5 μm operating at 1,800 K. Here the stopping field refers to the field at the facet center, which was calculated from the experimental extraction voltage and the gun geometry with a finite difference routine (SCWIM) in which the emitter tip end was approximated with a sphere, truncated by a facet with a diameter of $0.6r_{\text{tip}}$. With that field strength, and Eq. (5.6), the average surface tension for the Schottky emitter was calculated to be 1.4 J/m^2. This is to be compared to a surface tension of 2.8 J/m^2 for bare tungsten emitters (Bettler & Barnes, 1968). This surface tension gives for the minimum field to prevent collapsing of a Schottky emitter: $0.564/\sqrt{r_{\text{tip}}}$ (V/nm), with r_{tip} in μm. It is noted that Eq. (5.6) was derived for emitter profiles that have a specific ratio of the variation in surface curvature and electric field strength in the apex area, which does not necessarily apply to the (variable) shape of the Schottky emitter.

Dyke *et al.* (1960) have pointed out that the field strength of Eq. (5.6) may give a zero gradient in the apex area, but that elsewhere along the emitter surface the chemical potential still varies. In Fig. 5.12 we quantify this, for the Schottky emitter in Fig. 3.11 in the standard gun geometry, for a constant suppressor voltage of −300 V and for different extraction voltages with respect to the emitter. Figure 5.12 gives the (smoothed) chemical potential along the surface calculated for the first 20 μm starting at 100 nm from the facet edge. For the atomic volume that of tungsten was taken and for the surface tension 1.4 J/m^2 (Swanson & Schwind, 2008).

Figure 5.12 shows there is a gradient along the emitter surface for all voltages: there is no voltage for which the emitter shape is truly stable. It can thus be expected that the tip end itself may be stable temporarily but that the evolution will eventually resume, unless there are kinetic limitations.

For a zero field there is in Fig. 5.12 a continuous gradient pointing from the tip area to the shank, promoting net diffusion from the tip area to the shank: the tip size and the emitter recession as a function of time can be calculated from Eqs. (5.2)—see Swanson and Schwind (2008)—and (5.3), respectively. It can be seen that by applying a small field (2 kV) the tip–shank gradient becomes weaker.

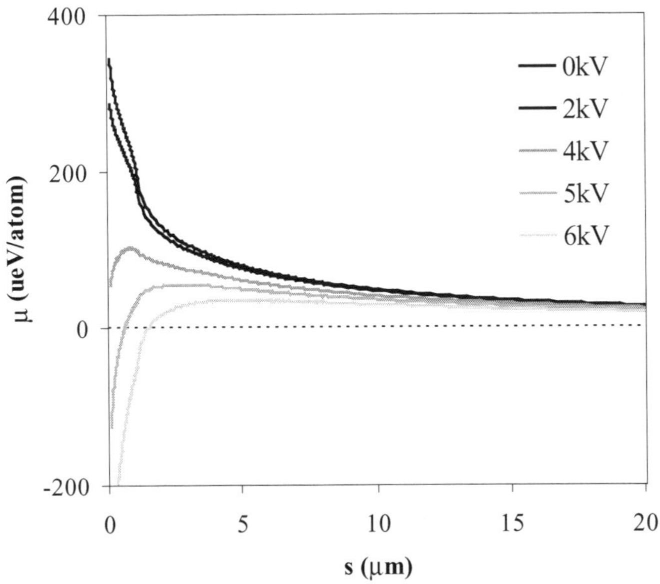

Figure 5.12 Estimated chemical potential along the emitter surface. The emitter facet is at $s = 0$ μm.

The minimum collapse prevention field, $0.564/\sqrt{r_{tip}}$ (Swanson & Schwind, 2008), for which there should be no gradient, at least not in the tip vicinity, depends on how r_{tip} should be interpreted for this specific emitter. For the tip radius as specified by the manufacturer for this emitter (0.85 μm) the stopping field would be 0.61 V/nm, and for a tip radius defined by the radius of the circle touching the end and side facets (Fig. 3.21), the stopping field would be 0.74 V/nm. For the standard gun geometry, with a constant suppressor voltage of –0.3 kV with respect to the emitter, this translates to an extraction voltage of around 4.0 kV to 4.5 kV. For this voltage range the gradient in the tip vicinity is indeed relatively weak, and also the chemical potential difference between tip and shank is small.

For larger extraction voltages (5–6 kV) the chemical potential at the tip becomes lower than that of the shank, promoting net diffusion from the shank toward the tip area. In other words, for fields above that in Eq. (5.6), Eq. (5.7) becomes an "extrusion" rate.

Note that for the higher voltages in Fig. 5.12 there is a maximum in the chemical potential a few microns away from the tip: the local gradient can be different from the gradient on a "global" scale (tip–shank). For example, for 4 kV the chemical potential at the tip is higher than at the shank, but in the tip area the local gradient actually points toward the apex.

The evolution of the geometry of the emitter in a field can be described with the continuum model above in a fashion similar as that was done by, for example, Nichols and Mullins (1965), if, for each time step, together with the change in geometry, the change in the field distribution along the surface is also evaluated. For larger fields, however, often the assumption of isotropic properties starts to break down, and what is usually seen in practice is not "extrusion," but so-called "build-up" (Barbour *et al.*, 1960).

5.3.2 Tip Size Growth at High Fields

With respect to build-up it is useful to consider the tip end as a small crystallite again, supported by the emitter cone. Contrary to a free crystallite, a change in the tip shape does not necessarily have to be realized with atoms from the tip area but can also be achieved by giving up atoms to the emitter cone or by adopting atoms from the cone.

It was shown in Fig. 5.12 that for larger extraction voltages the direction of net diffusion is toward the tip area, which may suggest emitter extrusion (as the opposite of the emitter shortening for net diffusion away from the tip area for low extraction voltage). But for large fields the "equilibrium" shape of the tip end becomes more faceted (see, e.g., Figs. 5.7 and 5.9), and emitter extrusion is prevented by the energy barrier involved in the outward expansion of the (end) facet. The energy barrier that prohibits expansion of the end facet outward is present because, as material is transferred onto the facet, the end facet must, at some point, be covered by a partial layer. This intermediate state has a high energy (Mullins & Rohrer, 2000).

As a consequence the tip size will grow due to net atomic transport from the shank toward the tip area (Fig. 5.12), and at the same time the tip shape becomes more faceted (Fig. 5.7), but the emitter does not become longer. (Note that with very high fields the nucleation barriers can disappear, because the surface tension can be zero or negative, and in such cases emitter extrusion would be possible—Fig. 5.10.)

We believe a nice example is the change of the emitter shape during the faceting routine employed in standard Schottky emitter production. Figure 5.13 shows a Schottky emitter before and after the routine that is used by the manufacturer to create the end facet, which involves applying a field to the emitter for the first time. The initial tip is spherical, while the end result shows a faceted geometry, and is also considerably larger. The change in tip size is ascribed predominantly to the mass transport from the shanks toward the tip, induced by the applied field. It is noted that a small part of the tip growth can in fact be attributed to material peeled off from the tip end through collapses. This will be discussed in more detail in Section 5.5.1.

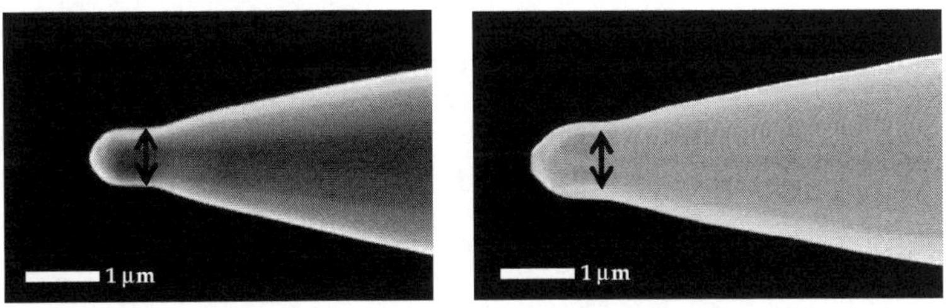

Figure 5.13 Schottky emitter before (left) and after (right) the faceting routine (stabilization at 0.5 mA/sr at 1,800 K). (SEM images from FEI Company.)

Once a tip is fully faceted, tip size growth might be slowed down due to the energy barriers involved in outward expansion of facets. The reshaping of the emitter just behind the tip will, however, continue due to the persisting chemical potential gradient pointing toward the tip. In support of this, analysis of profiles of Schottky emitters after production and after prolonged operation (e.g., Fig. 5.3) has yielded that in general at the end of the emitter life the cone angle in the tip vicinity is considerably lower.

Figure 5.14 is an illustration of the possible evolution for emitters with an already faceted tip end, subjected to a high field. The tip size grows due to net atom transport from the shank toward the tip, but the tip end maintains its faceted geometry.

It is noted that, for constant operating parameters, the change of the emitter profile will gradually reduce the field strength at the tip end. As a result the (equilibrium) shape of the emitter may become less faceted again over time and the chemical potential gradient reduces. Ultimately it may be that the direction of the chemical potential gradient reverses and the emitter could actually start collapsing.

To maintain the faceted geometry it may thus be required to increase the applied extraction voltage during the emitter lifetime.

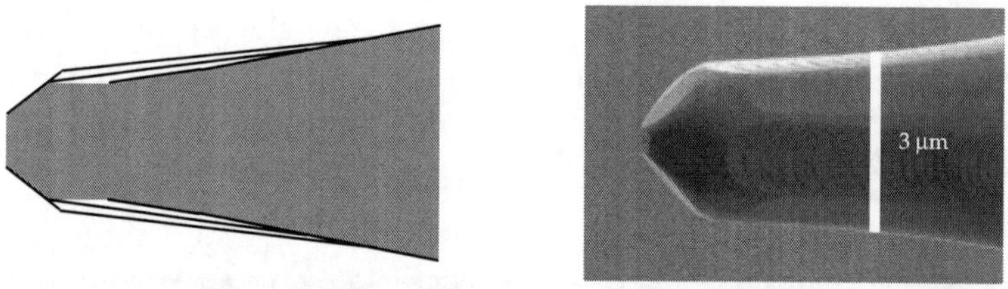

Figure 5.14 (Left) Illustration of the possible evolution for emitters with a faceted tip end subjected to a high field. The gray shape is the initial geometry. The black lines indicate the (outward) progression of the emitter contour for prolonged operation. (Right) SEM image of a Schottky emitter after serving 21 months in a commercial microscope. The tip size has increased considerably, but due to the fact that the system was used extensively without problems, it is not expected that the emitter has had many collapses of the end facet that would have disturbed the beam current. The extraction voltage has been increased two or three times during the emitter lifetime.

In conclusion, the tip size of standard Schottky emitters increases with increasing operation time, both at low and at high fields, and irreversibly.

5.4 Changes of the End Facet Geometry

We will now zoom in and concentrate on changes of the facet at the tip end. We will give two examples, one with the accent on the tip–emitter interplay and one with the accent on the almost(?) free crystallite behavior of the tip end—reversible changes.

We will use two experimental methods, monitoring of the facet emission pattern and monitoring of the Schottky plot slope (Section 3.4).

As discussed in Chapter 3, the facet emission pattern is a (distorted and blurred) "shadow" image of the facet, and the best "image" of the facet geometry is obtained at a relatively high extraction voltage and a relatively low temperature. This will be exploited below. Figure 5.15 is given as a reminder.

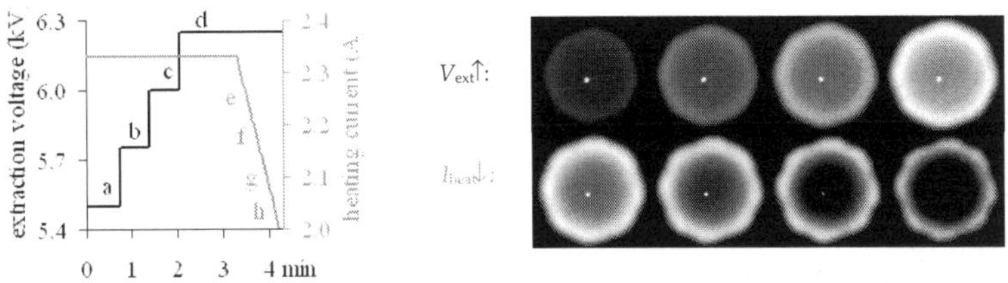

Figure 5.15 Effect of the operating conditions on the emission pattern. N.B. The central bright spot in the patterns is light emitted by the hot source, penetrating the YAG screen. All patterns are recorded with the same camera settings.

The Schottky plot slope is proportional to the square root of the field enhancement factor of the facet center, Eq. (3.14). And because the field enhancement factor of the facet center changes with the facet size (Fig. 3.25), it can be used to monitor changes.

5.4.1 Evidence of the Tip–Emitter Interplay

To illustrate the relevance of tip-emitter interplay we use experimental results obtained for the emitter in Fig. 5.16, operated in a setup as in Fig. 2.3 (setup B).

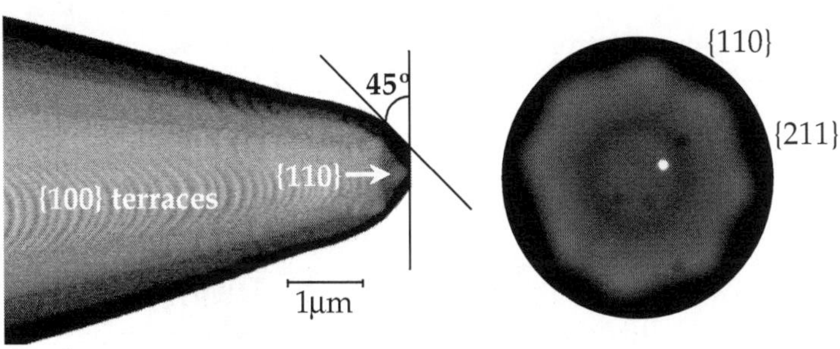

Figure 5.16 SEM image of a Schottky electron emitter as purchased and the emission pattern after starting up the source. The dark spots in the pattern are stains on the CCD camera lens. *Abbreviation*: CCD, charge-coupled device.

The emission pattern in Fig. 5.16, which is the emission pattern right after starting up the source, shows an octagonal facet shape with a dark ring. The octagonal shape of the emission pattern can be associated with the faceted tip-end geometry seen in the SEM image. The facets meeting the front facet correspond to the low index faces of the crystal—the {110} and {211} planes, which make angles with the {100} front facet of, respectively, 45° and 35°. The side-view SEM image suggests the edges of the {100} front facet touching the {110} facets are shorter than the edges touching the {211} facets, which would mean for the emission pattern in Fig. 5.16 the edges with the {110} facets are as indicated.

The dark ring indicates the presence of a circular {100} island on the facet. We don't expect this island to be formed during the start-up of the source, because the source temperature has not been high enough yet. We thus assume the island must have been present upon purchase and has only a small height, because it is not visible on the side-view SEM image in Fig. 5.16.

The faceted tip end is the result of the stabilization routine performed by the manufacturer (FEI Company) in the standard gun geometry at a final extraction voltage of 6.745 kV, a suppressor voltage of –240 V, and a temperature of 1,800 K. We translate the applied potentials to a field strength at the emitter surface with the charge density method and approximate the emitter in Fig. 5.16 with a rotationally symmetric version. At room temperature the emitter cone protruded 249 μm from the suppressor cap. For the simulation the tip protrusion is set at 0.28 mm to take into account thermal expansion. This gives a field strength at the facet center, associated with the stabilization conditions, of ~1.15 V/nm.

The stabilized geometry of Fig. 5.16 was started up to operate at low temperature (2.22 A, according to the manufacturer: 1,675 K), and at an extraction voltage of 5.2 kV and a suppressor voltage of –300 V. These potentials translate to a field at the tip of ~0.86 V/nm: lower than the field the tip was stabilized at. Figure 5.17 shows the initial emission pattern and the pattern after 64 hours of operation.

It can be seen in Fig. 5.17 that the facet becomes rounder and larger. This can be understood as a response to a change in the equilibrium crystal shape of the tip end, as illustrated in Fig. 5.18: upon decreasing the field strength the equilibrium shape looses the sharp edges, the size of the {110} facets decreases, but the size of the {100} facets increases.

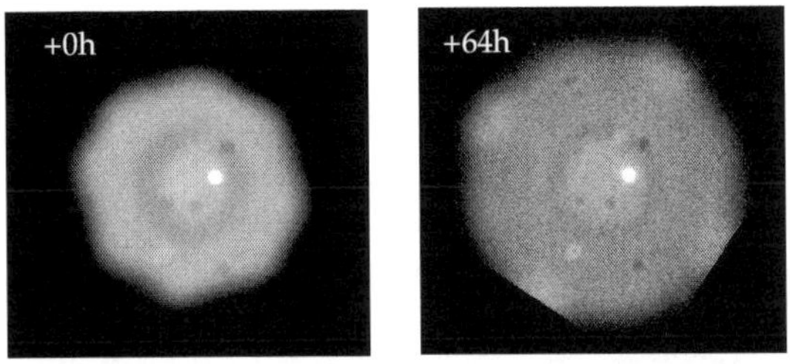

Figure 5.17 Change in the end facet geometry upon operating a stabilized polyhedral shape at a reduced field. The emission patterns are taken at equivalent extraction voltage and with a 25K emitter temperature difference. It is noted that the camera settings were slightly different. The difference in current in the patterns is <2%.

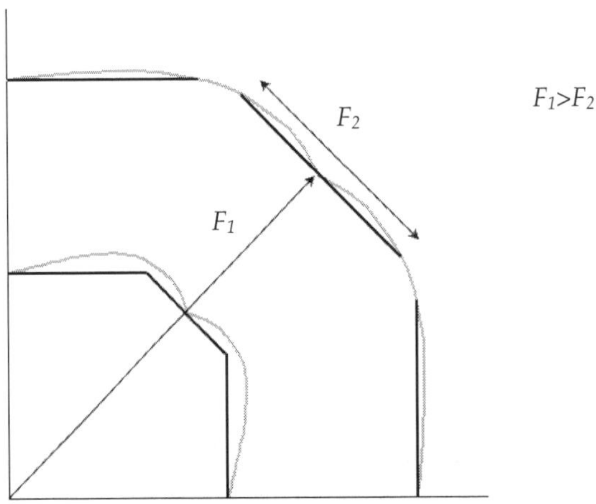

Figure 5.18 A Wulff plot with the same orientation as in Figs. 5.7 and 5.8, showing how a reduction in field strength could explain the experimental results in Fig. 5.17. Compare the size of the facets relative to the crystal size: for the lower field the {110} facets (horizontal and vertical facets) are smaller, but the {100} facet is actually larger.

For a free crystallite, for which the volume is conserved, the realization of the new equilibrium crystal shape would involve the collapse of the {100} facet, but this has not occurred in between the images in Fig. 5.17. It is believed that the change in tip-end geometry is assisted by the chemical potential gradient along the shank in the tip vicinity (Fig. 5.12), which, at 5.2 kV extraction voltage, is apparently still high enough to deliver atoms to the tip area (at least initially).

It is noted that upon further decreasing the field strength the equilibrium shape becomes more spherical, and the equilibrium {100} facet size will decrease again, and also the chemical potential gradient in the tip vicinity will change, yielding less atoms toward the tip or a change of the direction of the net diffusion. In the latter case the facet could start collapsing.

5.4.2 Reversible Changes of the End Facet*

Changes of the equilibrium shape of a crystallite induced by a change in the operating conditions are reversible. Here it will be shown that this also applies to the tip-end geometry of a Schottky emitter. In the first part we will use the evolution of the full emission pattern (Section 5.4.2.1) and in the second part the evolution of the Schottky plot slope for emission from the facet center (Section 5.4.2.2).

5.4.2.1 Monitoring with the emission pattern

The reversibility of changes of the end facet will be demonstrated with more results on the emitter in Fig. 5.16 operated in a setup as in Fig. 2.3 (setup B).

5.4.2.1.1 *Experimental*

The emitter was operated for slightly less than a month at different operating conditions. Figure 5.19 shows the applied conditions as a function of time. For clarity the short-term variations have been omitted in this figure.

The situation at the end of the third week is the starting point for the results that will be discussed here. In the final week the extraction voltage was increased for about a day (Section **A**), reduced again for several days (Section **B**), and finally increased again for almost two days (Section **C**). It was verified that all reported changes were achieved without a single collapse of the end facet.

For most of the time the emitter was operated at 2.27–2.33 A, which corresponds to a temperature range of 1,740–1,810 K in the test system of the manufacturer. The suppressor voltage was kept at −0.3 kV with respect to the emitter at all times. The background

*Most of this paragraph has been previously published and was reproduced with permission from *J. Vac. Sci. Technol. B*, **27** (2009) 2524, Copyright 2009, American Vacuum Society (see also page 249 of this book).

pressure in the system was ~1 × 10^{-7} Pa. In operation the pressure increased but was always below 1 × 10^{-6} Pa.

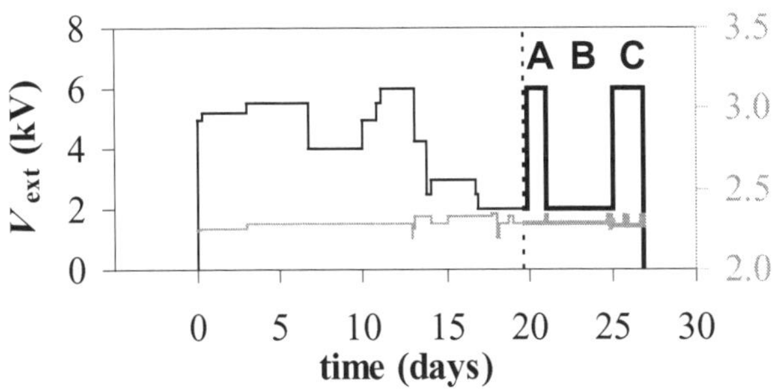

Figure 5.19 Operation conditions applied to the emitter during the experiments. Vertical dashed line marks the start of the experiments discussed here.

5.4.2.1.2 *Results*

The emission pattern at the end of the third week, a week in which the operating voltage was kept low, is the first pattern given in Fig. 5.20. The pattern is taken at an extraction voltage of 6 kV and a heating current of 2.29 A. Starting from this pattern, the operating voltage was kept at 6 kV for 27 hours. Figure 5.20 gives the evolution of the emission pattern.

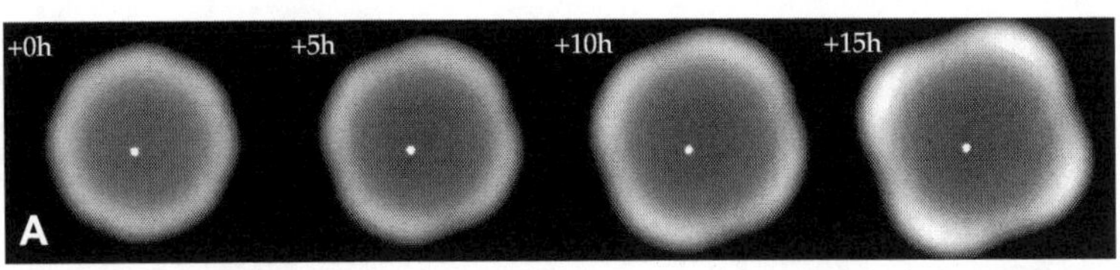

Figure 5.20 Emission patterns at 6 kV, 2.29 A, taken at 0 h, 5 h, 10 h, and 15 h after setting the extraction voltage to 6 kV (Section **A**).

It can be seen in Fig. 5.20 that upon increasing the extraction voltage to 6 kV after prolonged operation at low extraction voltage, the emission pattern starts to increase in size and changes in shape. Figure 5.21 gives the current in the pattern as a function of time. Upon increasing the extraction voltage the facet current increases from 26 μA to 43 μA (165%) in 15 hours' time.

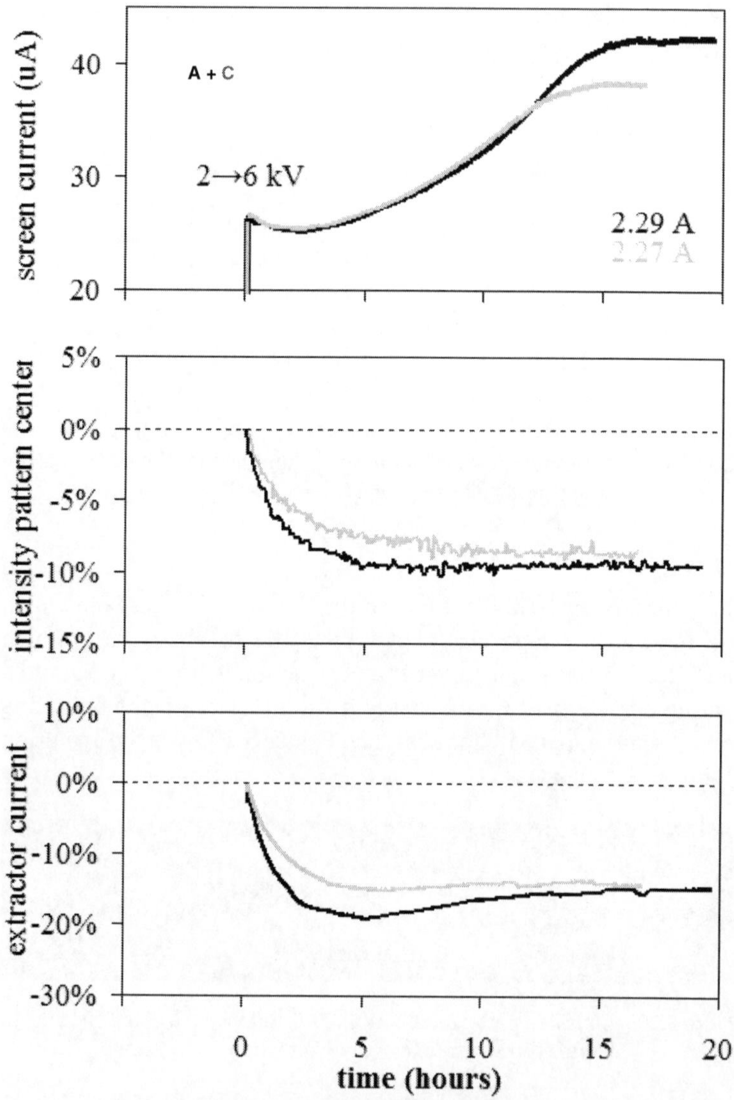

Figure 5.21 Response upon increasing the extraction voltage from 2 kV to 6 kV after prolonged operation at low voltage. Black curves correspond to Section **A** in Fig. 5.19 (patterns in Fig. 5.20), and gray curves correspond to Section **C** in Fig. 5.19 (patterns in Fig. 5.24). (Top) The current collected on the YAG screen. (Center) The relative change in the intensity of the pattern center. (Bottom) The variation in the current collected on the extractor.

After 17 hours the pattern shape and the current in the pattern appear to stabilize. Figure 5.21 also shows that while the total facet current increases, the intensity in the center of the pattern, relevant for (single-beam) applications, decreases. The third graph in Fig. 5.21 gives the extractor current profile. The current collected on the extractor at $t = 0$ is 71 μA, and 62 μA at 20 hours. The changes are significant, but its trend is currently not understood.

The extraction voltage was now reduced for several days (Section **B** in Fig. 5.19). Figure 5.22 shows two emission patterns recorded at 6 kV and 2.29 A: the first is taken just before reducing the extraction voltage and the second after 88 hours of operation at 2 kV.

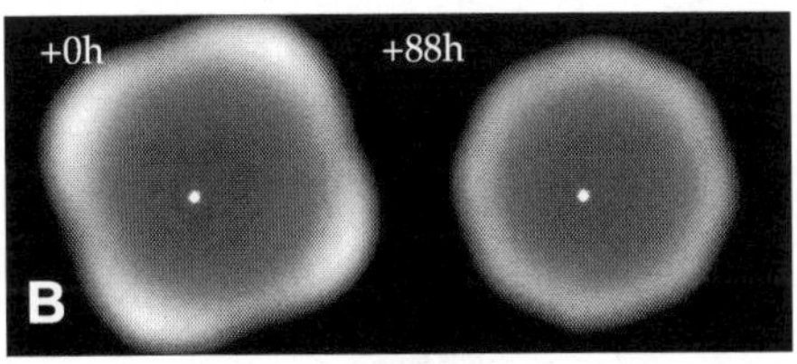

Figure 5.22 Emission patterns at 6 kV, 2.29 A (top) and low temperature (bottom). Right ones are after operating at 2 kV for 88 hours (Section **B**).

It can be seen that the pattern returns to a shape and size similar to the first pattern of Fig. 5.21. The current in the second pattern in Fig. 5.22 is 27 μA, compared to 26 μA in the similar-looking Fig. 5.20a pattern. Figure 5.23 gives the current collected on the screen, and the variation in the extractor current, during the operation at 2 kV. The screen current in Fig. 5.23 shows no sign of stabilization at 88 hours, and it is expected that for prolonged operation at 2 kV that the facet will eventually collapse. It is noted that the variation in the extractor current shows a trend opposite to that seen in Fig. 5.21.

Finally, the extraction voltage is increased again to 6 kV (Section **C** in Fig. 5.19). Fig. 5.24 shows the emission patterns at 0 hours, 5 hours, 10 hours, and 15 hours after setting the voltage back to 6 kV.

The screen current and the variation in the intensity at the pattern center and in the extractor current have been given in Fig. 5.21 in gray. The screen current increases from 27 μA to 38 μA (up by 141%). As before, the screen current seems to stabilize but this time after 15 hours. This was now further investigated. The emitter was operated at 6 kV for 43 hours in total, with occasional short-term variations. It was found that the facet current and the intensity in the pattern center for constant conditions remained constant within 1%.

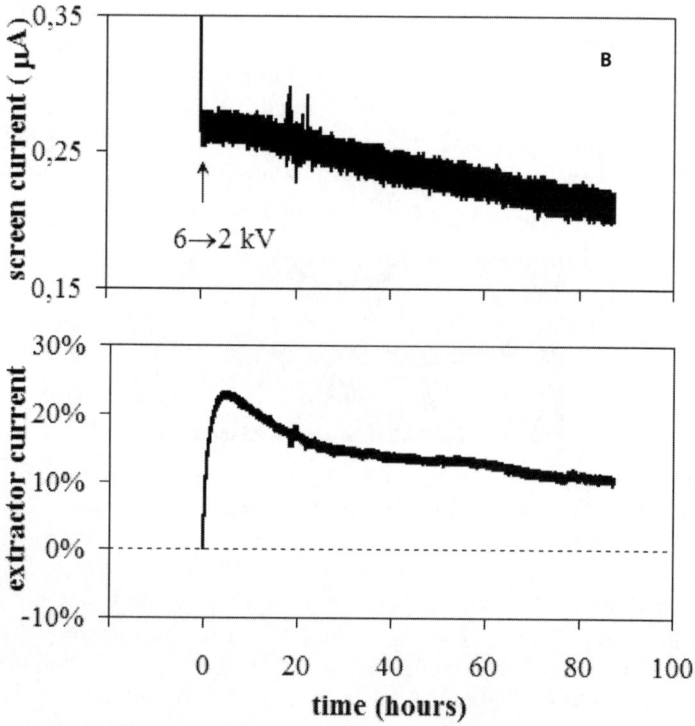

Figure 5.23 Response upon decreasing the extraction voltage from 6 kV to 2 kV. (Top) The current collected on the YAG screen. (Center) The relative change in the current collected on the extractor.

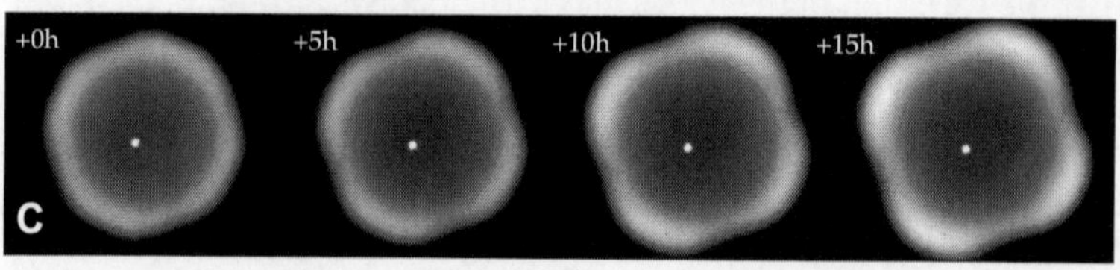

Figure 5.24 Emission patterns at 6 kV, 2.27 A, 0 h, 5 h, 10 h, and 15 h after changing the voltage from 2 kV to 6 kV (Section **C**).

As a check of the reproducibility the emitter was temporarily heated to 2.29 A and the pattern was compared to the stabilized pattern in Fig. 5.20. It was found that the size and geometry of the contour (same intensity contour line) were equivalent but that the current in the pattern taken first was a few percent higher.

To get a slightly sharper image of the facets the temperature has occasionally been reduced during the experiments. Figure 5.25 gives such cold patterns at the beginning and end of Sections **A**, **B**, and **C**.

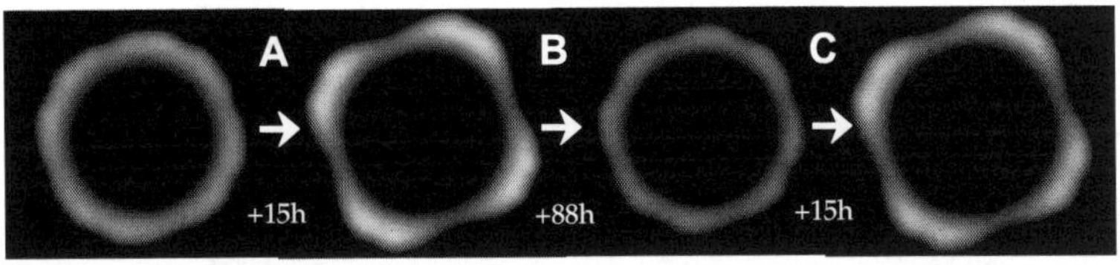

Figure 5.25 Emission patterns taken by temporarily reducing the heating current to ~1.97 A at a voltage of 6 kV after prolonged operation at low (patterns 1 and 3) or high extraction voltage (patterns 2 and 4) at normal operating temperatures.

Figure 5.25 shows that the pattern that develops both for low voltage operation (patterns 1 and 3) and for high voltage operation (patterns 2 and 4) is reproducible.

5.4.2.1.3 *Discussion*

For interpretation of the changes in the emission patterns in terms of the tip-end geometry, we first go back to the start of the experiments. The initial tip-end geometry was that in Fig. 5.16, which was the result of the stabilization routine performed by the manufacturer. The field strength associated with the stabilized shape was calculated to be ~1.15 V/nm. In this experiment the suppressor voltage was −300 V and the extraction voltage 2–6 kV, for which the associated field range can be calculated to be ~0.3–1.1 V/nm. The lowest field applied in the experiment is thus considerably lower than the stabilization conditions, while the highest field is close to the field associated with the stabilized geometry in Fig. 5.16. Note that the calculated fields should be considered approximates: the true tip shape is not rotationally symmetric, the thermal expansion is not known exactly, and it is expected that the emitter geometry as a whole evolves over time.

On the basis of patterns 2 and 4 in Fig. 5.25, and taking into account that the applied field strength is close the field strength associated with the shape in Fig. 5.16, it is believed that for prolonged operation at 6 kV the tip end is fully faceted, consisting of {211} and {110} planes connected to the octagonal {100} end facet, as it was upon purchase. A top-down view of the suggested tip-end geometry is given in Fig. 5.26 (right), in close agreement with the first SEM image of Fig. 5.4.

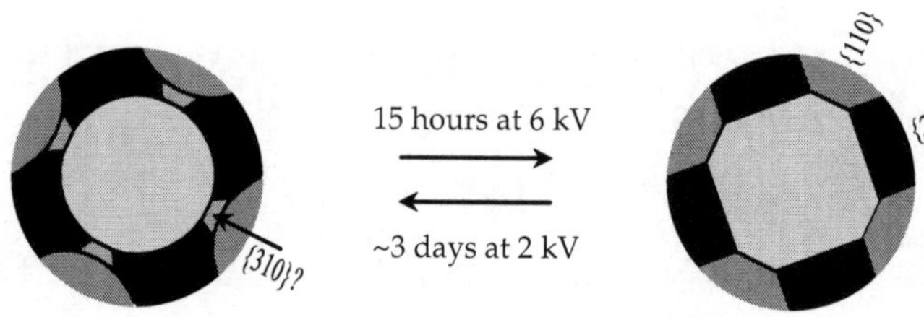

Figure 5.26 Top-down view of suggested tip-end geometry for prolonged operation at 2 kV (left) and 6 kV (right).

A fully faceted geometry could explain the fact that the pattern stabilized (at least temporarily): further increasing the end facet size would require (some of) the connecting end facets to expand outward, which is a process that generally involves a nucleation barrier. An example of this is believed to be the threshold voltage found by Fujita and Shimoyama (2007), which marked a sudden transition from a polyhedral tungsten tip end with {110} and {211} facets connecting to the {100} end facet to one with only {110} facets connecting to the {100} facet. A similar transition is expected for the Schottky emitter for operation at extraction voltages above 6 kV. And for operation at very high fields it is speculated that microprotrusions might form (Fujita & Shimoyama, 2007; Fursey, 2005).

Note that by changing the operating voltage from low to high, the net diffusion from tip to shank is reduced, stopped, or even reversed, depending on the variation in curvature and field strength along the surface. This will lead to a change in the geometry behind the tip end, which eventually could destabilize the tip end. Clear evidence of tip growth during the experiments was not observed, but an indication of tip growth, and thus a reduction of the average field enhancement factor, is the fact that the total facet current at a heating current of 2.29 A, after prolonged operation at 6 kV, is roughly 10% lower the second time (C compared to A).

For prolonged operation at low extraction voltage, the pattern is smaller than the patterns after operation at 6 kV (Fig. 5.25). The pattern shape is approximately circular, although there also might be facets connecting to the end facet, which could explain the four straight sections in between peaks of slightly increased intensity along the perimeter of the pattern. The smaller and more circular pattern geometry is in agreement with a more spherical tip end with smaller facets connected by rough regions, as expected for operation at low fields. The straight sections on the perimeter might be associated with the presence of {310} planes, which make a much smaller angle with the end facet than the {110} planes, but this cannot be confirmed. Note that the intensity along the perimeter of the pattern is more uniform in the patterns 1 and 3 than in patterns 2 and 4. The top-down view of a possible geometry has been given in Fig. 5.26.

For operation at low voltage the end facet is expected to collapse at some point, in response to the gradient in chemical potential that promotes net diffusion away from the sharp tip. The fact that there was no sign of stabilization is in agreement with this.

To get a better impression of how the tip end evolves, Fig. 5.27 gives overlays of contour lines of equivalent intensity, from the four images in Figs. 5.20 (Section **A**) and 5.24 (Section **C**).

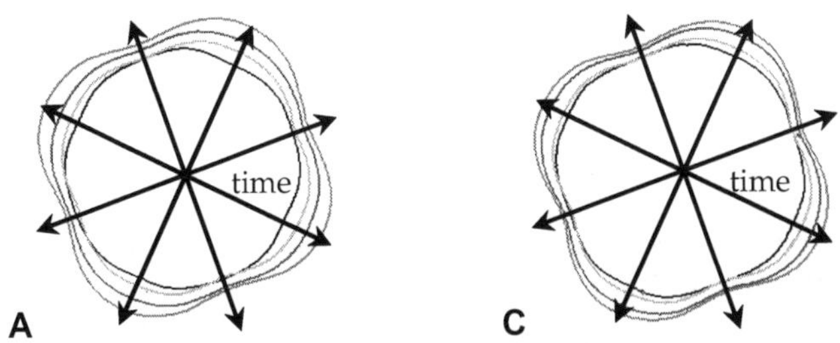

Figure 5.27 Overlays of the contours of constant intensity for the emission patterns in Figs. 5.20 (Section **A**) and 5.24 (Section **C**).

It can be seen that the pattern expands very little toward the {211} planes and much more toward the {110} planes. The expansion toward the {211} planes might be the result, not of a radial expansion of the end facet in that direction, but of sharpening of the facet edge: the rough sections are filled up, the {211} planes expand, and the ridges that separate the {211} planes from the end facet increase in length. This would make the emission pattern slightly larger (see Chapter 3). The expansion of the pattern toward the {110} planes is larger, and it is considered this is more likely to be the result of also some radial expansion of the facet, in addition to the effect of edge sharpening.

Changing from the low-voltage to the high-voltage pattern occurred much faster than changing from the high-voltage to the low-voltage pattern. This is not fully understood. Possibly this is related to the difference in the amount of rough area and the size of the facets present on the tip end between the low-voltage and the high-voltage shape, affecting diffusion rates, and/or the difference in the strength of the chemical potential gradient in the tip vicinity, between low- and high-voltage operation (Fig. 5.12).

5.4.2.2 Monitoring with the Schottky plot slope

A second method to monitor changes in the end facet geometry upon changing the operating conditions is by recording Schottky plots at regular time intervals. As discussed in Section 3.4 the slope of a Schottky plot is proportional to the square root of the field enhancement factor for the area that delivers the measured electrons. Changes of the facet size change the field enhancement at the facet center (see Section 3.3.1), which can thus be recorded by monitoring the Schottky plot slope for emission from the facet center as a function of time. Figure 5.28 gives an example of the change in the Schottky plot slope squared as a function of time for changes in the operating extraction voltage as indicated for a standard Schottky emitter in the standard gun geometry. The experimental setup is similar to that in Fig. 2.3 (setup B), but instead of a YAG screen there is a screen with an aperture. The current through the aperture is measured and converted in an angular intensity. It is noted that the data have been taken automatically for a fixed set of angular intensities by changing the extraction voltage and keeping the suppressor voltage constant. Because of this, the Schottky plot slope cannot give the true field enhancement factor, as the field form around the emitter has not been kept constant. But the trend in the slope is still a useful indicator of a change in the emitter geometry.

As can be seen the Schottky plot slope slowly increases when the extraction voltage is reduced and rapidly decreases again when the original extraction voltage is restored. This is in agreement with a facet size reduction for an increase in voltage and a facet size growth for a reduction in voltage (see Fig. 3.25) in agreement with equilibrium crystal shape theory (Figs. 5.7 and 5.8). The difference in the rate of change for increasing or decreasing voltage is in agreement with the result in the previous section.

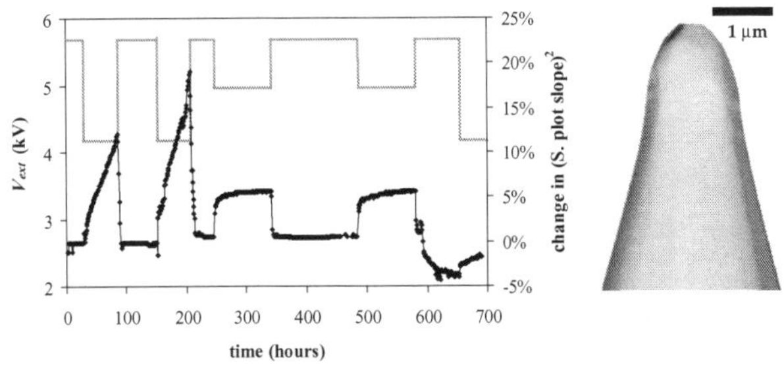

Figure 5.28 Variation in the Schottky plot slope upon changing the extraction voltage. The temperature was kept constant up to 770 hours, when it was decreased.

It is noted that for a small reduction of voltage (at ~350 h and ~480 h) the slope seems to stabilize at a new value, but for a large reduction of voltage (at ~40 h and ~160 h) the slope shows no sign of stabilization. It is might be that for the latter case the facet of the emitter will collapse if the voltage is kept low for a longer time period, in response to the chemical potential gradient, while for the small voltage reduction the facet stabilizes for a smaller diameter, which is associated with the crystal equilibrium shape at that voltage.

The collapsing of Schottky emitters will be discussed in more detail in the next section.

5.5 Collapsing of the End Facet*

The continuum theory of Section 5.3.1 can explain the observed recession and tip growth, but it cannot describe the phenomenon of the collapsing rings itself.

Over the last two decades, the interest in the stability of nanostructures has increased, and collapsing rings are now recognized as a decay mechanism that can be adequately described with step-flow models (Uwaha, 1988; Israeli & Kandel, 1999; Degawa *et al.*, 2006a). Step models can also deal with the presence of the end facet, which was ignored in the continuum approach. In step models

*Most of this paragraph has been previously published and was reproduced with permission from *J. Vac. Sci. Technol. B*, **26**, pp. 2073–2081, Copyright (2008), American Vacuum Society, and reprinted from *Ultramicroscopy*, **110**, pp. 1243–1254, Copyright (2010), with permission from Elsevier (see also page 249 of this book).

the surface area near a low index orientation is described in terms of an array of atomic steps, or a stack of terraces of atomic height, which evolves by net atom transport between steps. Step models have been used to describe the decay of (nano)structures on surfaces but also to describe how the facet size can increase on otherwise spherical crystallites upon reducing temperature (Degawa *et al.*, 2006b).

5.5.1 The Step-Flow Model

To explain the phenomenon of collapsing rings on emitters we follow Israeli and Kandel (1999) and model the very tip end of the emitter as a stack of (100) terraces bound by circular concentric steps (Fig. 5.29).

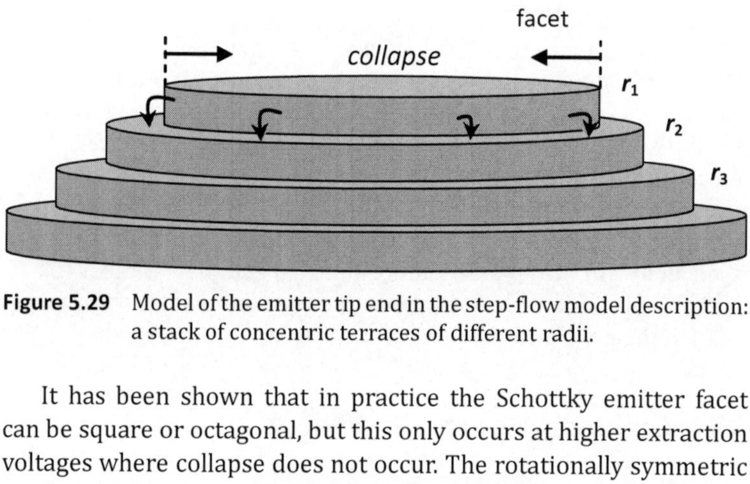

Figure 5.29 Model of the emitter tip end in the step-flow model description: a stack of concentric terraces of different radii.

It has been shown that in practice the Schottky emitter facet can be square or octagonal, but this only occurs at higher extraction voltages where collapse does not occur. The rotationally symmetric stack is in agreement with SEM images of Schottky emitters frozen in the collapse process, such as in Fig. 5.2.

A collapse is the shrinking of the top terrace or the top bunch of terraces. The terrace (bunch) shrinks when its bounding step emits atoms onto the underlying terrace. These adatoms can diffuse across the surface and can be absorbed at the steps bounding the terrace. To describe the collapse we look at atom transport from and to the steps. Evaporation, bulk diffusion, and formation of islands and voids are ignored. To solve the diffusion equations we need the equilibrium concentrations of adatoms on the terraces. The local equilibrium adatom concentration on a terrace depends on the work to create the adatom and the temperature (Israeli & Kandel, 1999):

$$C^{\text{eq}} = \frac{1}{\Omega_{\text{s}}} e^{-U_{\text{ad}}/k_{\text{B}}T}, \tag{5.8}$$

with atomic area Ω_s, adatom formation energy U_{ad}, the Boltzmann constant k_B, and temperature T. The adatom creation energy U_{ad} is defined as the work to bring an atom from a kink site in a step onto the terrace. The work to create an adatom depends on the local environment, such as the local curvature of the step and the local step density. Because the terraces in Fig. 5.29 are not equivalent this gives rise to differences in concentrations across the stack. If temperature allows diffusion will take place and the steps will start moving due to a net adatom absorption or emission. The equilibrium adatom concentration C_i^{eq}, near the curved step edge of terrace i with radius r_i, can be written as (Israeli & Kandel, 1999)

$$C_{\text{i}}^{\text{eq}} = \frac{1}{\Omega_{\text{s}}} e^{\left(-U_{\text{ad0}} + \frac{\Omega_{\text{s}}\beta}{r_{\text{i}}} + \sqrt{\Omega_{\text{s}}} \frac{\partial\left[U_{\text{int}}(r_{\text{i}}, r_{\text{i+1}}) + U_{\text{int}}(r_{\text{i}}, r_{\text{i-1}})\right]}{\partial r_{\text{i}}}\right) \Big/ k_{\text{B}}T}, \tag{5.9}$$

with β the step line tension, being the work per unit length to extend the step, and U_{int} the interaction energy between two steps.

The $1/r$ term gives a reduction of the work to create the adatom due to the step curvature. As a result the equilibrium adatom concentration is larger for terraces of smaller radius. Net diffusion will take place from the small terraces toward the lower terraces, and the stack will decay.

Step–step interaction can be either attractive or repulsive, so it can either decrease or increase the work to create an adatom. There is a general consensus that the most important step interaction between steps with step separation Δr corresponds to a $1/(\Delta r)^2$ interaction law (Bonzel, 2003), which covers three contributions: entropic interaction due to the formation of kinks and dipole–dipole interactions due to elastic as well as electric dipoles associated with steps. The $1/(\Delta r)^2$ interaction law for steps on a rotationally symmetric structure, as in Fig. 5.29, can be written as (Tanaka *et al.*, 1997)

$$U_{\text{int}}(r_{\text{i}}, r_{\text{i+1}}) = 2\pi g \frac{r_{\text{i}} r_{\text{i+1}}}{r_{\text{i}} + r_{\text{i+1}}} \frac{1}{(r_{\text{i}} - r_{\text{i+1}})^2}, \tag{5.10}$$

in which g is a measure of the interaction strength. For the cone-shaped stack of Fig. 5.29, this model gives an interaction contribution in Eq. (5.12) that decreases with increasing terrace radius.

The geometry of the stack yields an adatom concentration gradient, but how the system responds to this depends on how fast the adatoms diffuse across the terraces and across the steps. We can calculate the evolution of the tip end by solving the diffusion equation for adatoms on the terraces between the steps, with appropriate boundary conditions applied at the step edges. This gives a set of coupled equations of motion for the steps (Israeli & Kandel, 1999). Usually two cases are distinguished—slow terrace diffusion and fast attachment and detachment from the steps versus fast terrace diffusion and slow attachment and detachment from the steps. With strong step–step repulsion, large g in Eq. (5.10), Israeli and Kandel (1999) have predicted for both cases the successive collapse of single atomic layers, while for weak step–step repulsion they showed that in the second case the steps can bunch together and collapse collectively. Then the collapses of lower terraces overtake the collapse of a higher terrace and the collapse continues as a many-layer event. Both single-layer collapse and step bunch formation and collapse have been observed experimentally for different materials.

Single-atomic-layer collapses have, for example, been reported for W(110), Mo(110), Ta(110) (Bettler *et al.*, 1974), Re(10-10), Re(0001), Ro(111), Ir(111) (Bettler & Barnes, 1968), Pb(111) (Degawa *et al.*, 2006b), Cu(111), (Gie01), and Au(100) (Hirai *et al.*, 2001) structures, while step bunch formation and collapse was noticed, for example, for W(110) and Mo(110) emitters in the presence of carbon or silicon (Bettler *et al.*, 1974). It is known that adsorbates can induce step bunch formation (Horn-Von Hoegen *et al.*, 1998), and considering the height of islands on the facet in a collapse process (Fig. 5.2) this also seems to be the case for the Schottky emitter, which has the ZrO_x adsorbates on the W(100) terraces. It is noted that for clean W(100) single-atomic-layer collapses have been reported (Bettler & Barnes, 1968).

To take into account the effect of the extraction voltage we follow Giesen *et al.* (2005). Key to this approach is the change in the local charge density at the surface upon the creation of an adatom (Müller, 2006) (Fig. 5.30). The charge redistribution induced by an adatom can be characterized with a dipole moment p and changes the work function of the surface

$$\Delta V = \frac{Cp}{\varepsilon_0}, \tag{5.11}$$

with C the local adatom concentration.

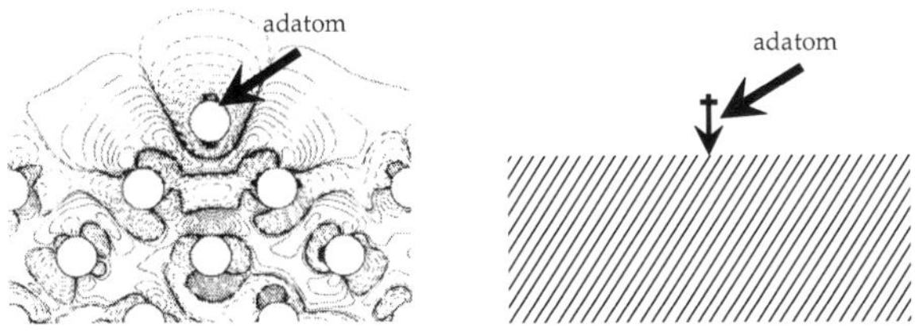

Figure 5.30 An adatom on the surface changes the charge distribution. The left image is taken from Müller (2006) and shows the change in the charge density induced in the (110) plane by the adsorption of the adatom on Au(100) from ab initio calculations. The continuous and dashed contour lines represent an increase and a decrease of charge. The right image is a schematic representation of the adatom as a dipole on the surface.

For the emitter in operation the work function change means a change of the potential difference between the emitter and the extractor, and thus a change in the work the power supply has to do to bring back the extraction potential to its original value. Each adatom is thus accompanied with a compensating charge flow. The compensating work per adatom delivered by the power supply to restore the extraction potential equals

$$-\frac{\sigma\Delta V}{C}=-\frac{\sigma p}{\varepsilon_0}=-pF\,, \tag{5.12}$$

with σ the local charge density (negative field points away from surface). The total work to create the adatom on the surface of an emitter in operation is thus different from the work at zero extraction potential:

$$U_{\mathrm{ad}}=U_{\mathrm{ad0}}-pF \tag{5.13}$$

A positive dipole moment would increase the work per adatom for an operating emitter; a negative dipole moment would decrease it. This means that when the extraction voltage is applied Eq. (5.9) changes to

$$C_{\mathrm{i}}^{\mathrm{eq}}=\frac{1}{\Omega_{\mathrm{s}}}e^{\left(-U_{\mathrm{ad0}}+pF+\frac{\Omega_{\mathrm{s}}\beta}{r_{\mathrm{i}}}+\sqrt{\Omega_{\mathrm{s}}}\frac{\partial\left[U(r_{\mathrm{i}},r_{\mathrm{i+1}})+U(r_{\mathrm{i}},r_{\mathrm{i-1}})\right]}{\partial r_{\mathrm{i}}}\right)\Big/k_{\mathrm{B}}T}. \tag{5.14}$$

Steps also have dipole moments, and the step line tension and step–step interaction will also change with the applied potential difference (Ibach & Schmickler, 2003), but we will neglect this.

The field dependence of the equilibrium concentration will affect the concentration on the terraces of the stack in Fig. 5.29. For a positive adatom dipole moment the equilibrium adatom concentrations will decrease with increasing extraction potential (increasingly negatively charged).

Due to the geometry of the emitter the charge density on the emitter surface is a function of position and has thus a position-dependent contribution to the equilibrium adatom concentration. The pF term in Eq. (5.14) changes to $pF(s)$, where s is the distance along the surface measured from the facet edge. The relation between s and the step curvature r depends on the tip geometry. Decay on emitter tip ends can thus, in principle, be prevented by an electric field gradient that balances the step curvature term.

5.5.1.1 Application to Schottky emitters

We apply the theory above to the emitter tip profile shown in Fig. 5.31. We consider three different tip-end geometries and look at the first tens of nanometers below the facet, because this is the area relevant to step bunches on the tip of Schottky emitters (Fig. 5.2).

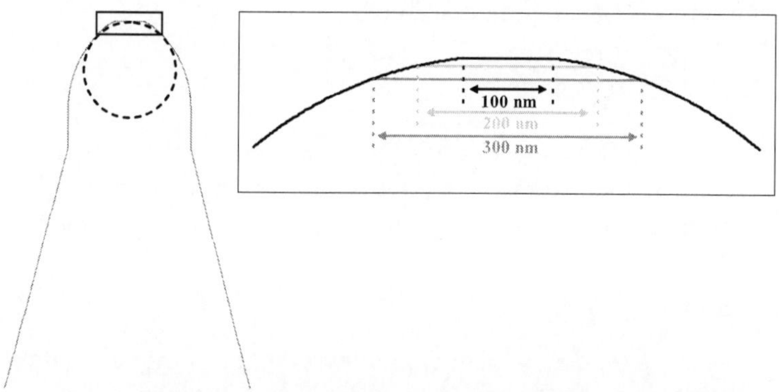

Figure 5.31 Profile of the Schottky emitter used in field calculations and a close-up of the tip-end geometries investigated.

The field distribution along the tips is calculated for different extraction voltages with the charge density method (http://www.electronoptics.com) for operation in the standard gun geometry. The distance between the facet and the extractor is slightly different for the different tip-end geometries (Fig. 5.31), but its effect on the field at the emitter surface is negligible. Figure 5.32 gives the calculated field strength across the emitter surface (at a 6 nm distance) for an extraction voltage of 4 kV.

Figure 5.32 shows that the electric field strength has a peak near the facet edge, which increases with increasing facet diameter. This is related to the increasing contact angle (the angle at the facet edge between the plane of the facet and the tangent to the curved surface) with increasing facet diameter (here for the 100 nm facet: 7º; 200 nm facet: 13º; 300 nm facet: 20º). The field at the facet center that is relevant to the probe quality decreases with increasing facet diameter. The portion relevant to collapse prevention in our model is the part of the curve that starts at the facet edge. For that part the field and the field gradient are strongest for the largest facet.

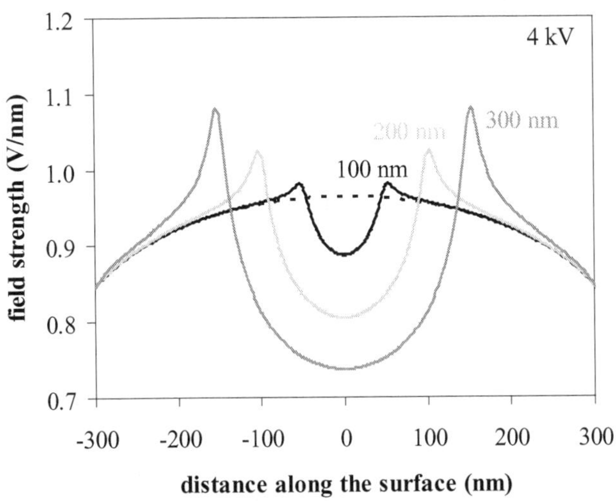

Figure 5.32 Electric field strength along the surface of a Schottky electron emitter (Fig. 5.31) in a standard gun module with a suppressor voltage of −0.3 KV and an extraction voltage of 4 kV. The dashed line indicates the field distribution without a facet.

To calculate the effect of these field distributions on the equilibrium concentrations near the steps several material properties are required—Eq. (5.14). The atomic area of W(100) is 1.00^{-19} m^2, but for the Schottky emitter the other parameters in Eq. (5.14) are basically unknown. In first approximation we neglect step–step interaction and the effect of the extraction voltage on the step line tension. How the zirconium and oxygen affect the step properties is not known, but it can be shown that for a step dipole moment of 0.01 eÅ (step dipole moment for a step in the <110> direction for a W{001} plane [Gardiner *et al.*, 1981]) the field dependence of the step line tension is a second-order effect. To illustrate the effect we are mainly interested in the step curvature term and the field term in the exponent of Eq. (5.14). Therefore we rewrite Eq. (5.14) as

$$C^{eq}(s) = C_0 A(s) = C_0 e^{\frac{\Omega_s \beta / r(s) + pF(s)}{k_B T}}, \tag{5.15}$$

with C_0 the equilibrium adatom concentration for a large terrace in the absence of a field—Eq. (5.8) with $U_{ad} = U_{ad0}$—and calculate only the multiplication factor $A(s)$. For the multiplier A we only need a step line tension and an adatom dipole moment. For the adatom dipole moment we use 0.05 eÅ. This is less than the 0.2 eÅ for a tungsten adatom on the most densely packed tungsten plane (110), reported by Besocke and Wagner (1973), and also smaller than the ~0.1 eÅ calculated by Müller and Ibach (2006) for the (100) planes of the face-centered cubic (fcc) materials Au, Cu, and Ag, which also have a higher packing density than the (100) plane of the body-centered cubic (bcc) structure of tungsten. For W(110) step line tension data are available (Xu *et al.*, 1994) and range from 0.1 eV/Å to 0.3 eV/Å, depending on the step orientation. Rather arbitrarily we use for our system the average value of 0.2 eV/Å. The temperature is the standard operating temperature for Schottky emitters is 1,800 K.

Figure 5.33 shows how multiplier A varies along the surface of the different tip ends in Fig. 5.31 in the absence of a field and for the field distribution shown in Fig. 5.32. A larger A value means a higher equilibrium adatom concentration at that position.

At 0 kV the equilibrium adatom concentration decreases with increasing distance from the facet edge because the terrace radius increases going in this direction. Atom transport points away from the apex, and it is expected the facet will collapse. The tip end with the smallest facet has the largest concentration and largest concentration gradient.

By applying the extraction voltage the concentration reduces because the positive dipole moment increases the work to create an adatom. The effect of the voltage is smallest for the tip with the 100 nm facet because the field and field gradient are smallest (Fig. 5.32). For this facet the concentration gradient still is such that atom transport is pointing away from the apex, and this is also the case for the 200 nm facet tip. For the tip with the 300 nm facet, however, for which the field and field gradient are strongest, the concentration gradient is reversed. For this tip the atom transport is expected to be toward the facet. For the chosen material properties and extraction voltage collapsing rings are thus expected for the 100 nm and 200 nm facet tips but not for the 300 nm facet tip. From Fig. 5.33 we conclude that for a positive dipole moment the driving force for decay through collapsing rings due to step radius differences will be reduced and can even be reversed when the potential difference between the emitter and the extractor is applied. We think this applies to the experimental "ring-stopping field" results.

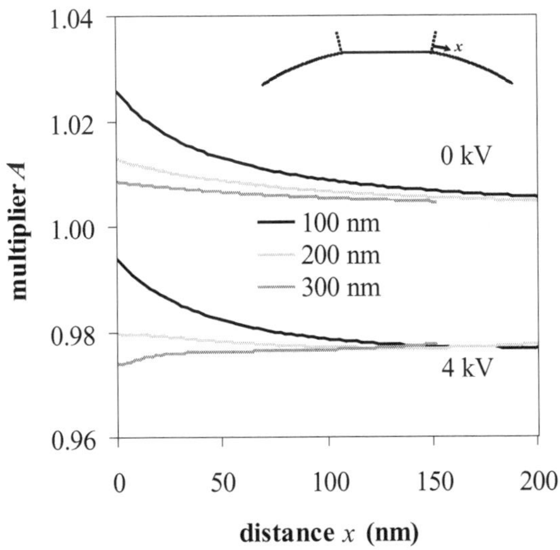

Figure 5.33 Variation of multiplier A, proportional to the adatom equilibrium concentration, Eq. (5.15), along the surface of the emitters in Fig. 5.30, with p = 0.05 eÅ, β = 0.2 eV/Å, and T = 1,800 K.

Whether or not a given emitter tip is expected to collapse and cause probe instabilities for a chosen extraction voltage thus depends on the exact emitter geometry and the field distribution along its surface. From Eq. (5.15) the condition for collapse prevention is

$$\frac{\mathrm{d}F(s)}{\mathrm{d}s} \geq \frac{\Omega_s \beta}{p} \frac{\mathrm{d}r^{-1}(s)}{\mathrm{d}s}, \tag{5.16}$$

which has to be valid along all of the stack.

For the tip-end geometries investigated here and the extraction voltage of 4 kV this is true only for the 300 nm facet tip. The tips with the smaller facets require much larger, impractical extraction voltages (for a tip with a 200 nm facet, voltage > 7 kV): for these tips a collapse cannot be prevented.

It is noted that for the investigated shapes no extraction voltage could be found for which A was constant along the whole tip-end surface and the tip would be fully stable. Figure 5.34 shows the variation of the multiplier along the surface of the tip with the 300 nm facet for different extraction voltages. The most stable probe from the emitter is expected for a condition in which the variation of A is small and the direction of net diffusion is not pointing away from the apex anywhere on the tip end (no collapse). From Figs. 5.33 and 5.34 this would be for the tip with the 300 nm facet an extraction voltage between 3 kV and 4 kV.

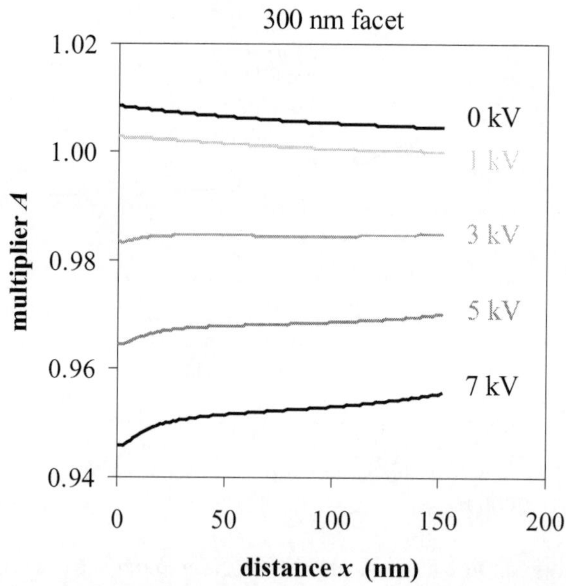

Figure 5.34 Variation of multiplier A, Eq. (5.15), along the surface of the emitter with a facet of 300 nm in diameter (Fig. 5.14) as a function of extraction voltage, with $p = 0.05$ eÅ, $\beta = 0.2$ eV/Å, and $T = 1{,}800$ K.

To make a comparison with the experimental ring-stopping field found by Swanson and Schwind (2008) we would have to know the exact shape of the tip end for which that result was obtained. The conversion of their experimental extraction voltage to a ring-stopping facet field was based on finite-difference calculations for a sphere-on-cone geometry, with the sphere truncated by a facet with a diameter of $0.6r_{tip}$. For an emitter with a 0.5 μm tip radius this gives a facet diameter of 300 nm and a contact angle of 17º. Our emitter profile (taken from an SEM image contour) is slightly different. We have drawn a circle of 0.5 μm radius in the profile in Fig. 5.31 for comparison. The field at the facet center of the tip with the 300 nm facet for which collapse is not expected is somewhere between 0.53 V/nm and 0.74 V/nm (corresponding to 3–4 kV extraction voltage) to compare to the experimental ring-stopping field of 0.8 V/nm (Swanson & Schwind, 2008). It is noted that the position and shape of the curves in Figs. 5.33 and 5.34 will change with different estimates of the dipole moment and step line tension and by taking into account step–step interaction and the field dependence of the step properties. Altogether we believe our results to be in reasonable agreement with the experimental value given by Swanson and Schwind (2008).

5.5.1.2 Discussion

It is noted that the model can only give the momentarily expected direction of diffusion. To calculate the evolution of the tip end the diffusion equations have to be solved, which requires a few more material properties, and an extension of the model to the shanks of the emitter, because we need a boundary condition of zero transport. We do not expect our model (based on a stack of {100} terraces) to hold for the range of crystallographic orientations traversed when going from tip to shank.

Also, the step model description is limited to rotationally symmetric tip ends. For a faceted tip the end facet will no longer be circular. Its edges will match directly to other low-index facets, and the tip-end area can no longer be described with a stack of concentric {100} terraces.

From Figs. 5.33 and 5.34 it can be seen that by applying an extraction voltage the value of A can become smaller than 1. A value of 1 is equivalent to the equilibrium concentration on a large terrace with a zero field. The electric field strength decreases with increasing distance from the apex, which implies that for {100} terraces further down the shanks of the emitter the multiplier value is ≥ 1. By applying an extraction voltage the concentration difference between the tip area and the shank area is thus also changed, which could give net diffusion toward the tip. This suggests that depending on the tip geometry and extraction voltage the concentration gradient in the tip area can point away from the tip end (collapsing rings), while at the same time there is a concentration difference just behind the tip area, promoting diffusion toward the tip (e.g., for 4 kV for the 100–200 nm facet tip in Fig. 5.33). This was also shown in Fig. 5.12.

On the basis of Fig. 5.33 we concluded that for the more spherical tips (with no or small facets) collapsing basically cannot be prevented. An illustration of this is believed to be the evolution of an emitter tip during the faceting routine that is employed by the emitter manufacturer to create the end facet. Figure 5.13 had shown that at the beginning of the faceting routine the tip end is spherical, which changes to a more faceted and larger shape as a result of applying the field. In Section 5.3.2 the bulk of the tip size growth was associated with net atom transport from the shank toward the tip, in combination with the energy barriers to expand facets outward. It is known, however, that the faceting routine involves at least several collapses (Swanson & Schwind, 2008).

The step-flow model predicts that the facet size will grow with every collapse (Israeli & Kandel, 1999). This is expected to apply also in the presence of a field, provided the tip is close to spherical, because then the field gradient along the surface will be much smaller than the gradient in step curvature. This describes the situation in the beginning of the faceting routine: the right-hand-side term in Eq. (5.16) is much larger than the left-hand-side term, and the emitter will start collapsing. With increasing facet size, however, the step curvature gradient decreases, right-hand side of Eq. (5.16), while the field gradient will increase (Fig. 5.32), left-hand side of Eq. (5.16). At some point the field term will start exceeding the curvature term, and the collapsing will stop: the end facet has formed.

Although the exact operating procedure for facet making is unknown, we use Fig. 5.35 to illustrate approximately what happens with the tip end during the faceting routine. It shows the evolution of the emission pattern for an emitter at 1,750 K after the field has been off for 20 hours.

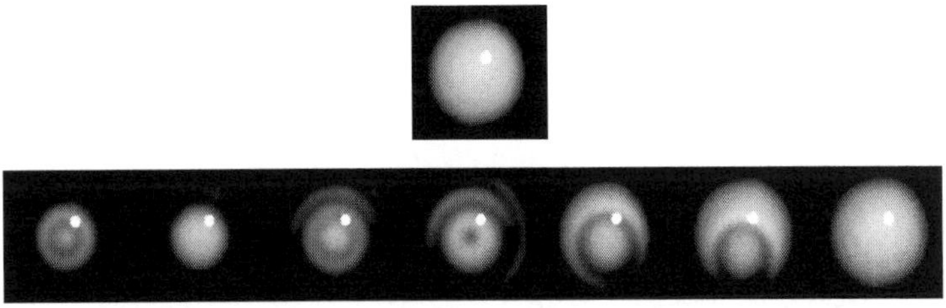

Figure 5.35 (Top) Emission pattern before the voltage is turned off. (Bottom) Snap shots of the evolution of the emission pattern after 20 h without a field. In 12 h the end facet is restored. The operating temperature is 1,750 K.

Although not exactly the same as the faceting routine, in which one starts with a spherical emitter that has not been in a field yet, the 20 hours without a field did change the tip end into a more spherical shape with the ringed emission pattern (Swanson & Schwind, 2008) typical for a collapsing and receding emitter at elevated temperature. The bottom half of Fig. 5.35 shows that upon restoring the field the ringed pattern changes (back) to that associated with a single facet, typical for operation in the presence of an extraction voltage that gives a field strength at or above the stopping field. It is noted that the facet of the last image remained stable for at least several days (until the conditions were changed again).

5.5.2 Tip–Emitter Interplay

For collapsing emitters there is the continuous gradient in chemical potential between the tip and the shank, yielding net atom transport away from the tip area, but it has been shown that nevertheless the tip geometry responds to changes in the operating conditions.

Experimental results upon cooling collapsing tungsten emitters suggested that the size of the collapsing facet gradually increased toward a new steady-state value, while upon heating a collapsing emitter, it was found that the collapsing temporarily halted (Bettler & Love, 1967). And a similar effect was observed for a change in the applied (low) bias (Barbour *et al.*, 1960). The response of the collapsing emitter upon a reduction in temperature or bias increase can be interpreted as "relaxation" of the tip end toward its new equilibrium shape, which has a larger facet. The temporary halt of the collapsing upon a temperature increase or bias reduction can be connected with the energy barrier associated with the nucleation of a new layer on the facet (Mullins & Rohrer, 2000; Degawa *et al.*, 2006a; Thürmer *et al.*, 2003). Instead of nucleating a new smaller-diameter layer on the existing end facet, the tip end reshapes to get a smaller facet by transporting the atoms from the facet perimeter away from the tip toward the shank, along the chemical potential gradient, and then resumes the collapsing.

The interpretation of these results is facilitated by the assumption that the collapsing does not affect the field at the tip nor the gradient in chemical potential between the tip and the shank, at least not for tens of collapses in a row. This is because the change in tip size (tip size is of the order of a micron) per collapse (atomic layer) is very small.

This does not apply to collapsing on Schottky emitters, for which a collapse involves many atomic layers. A single collapse can change both the local situation at the tip and the global gradient in chemical potential, and this complicates the analysis, as will be shown below.

Before we continue we first define one "ring series" as one period of beam fluctuations, as shown in Fig. 5.5, which starts and ends with the standard tip geometry that has a single end facet, as in, for example, Fig. 5.36.

5.5.2.1 Experiments

The experiments have been performed with a commercial Schottky source supplied with a suppressor and a heating wire (FEI Company). An SEM image of the emitter before the experiments is given in Fig. 5.36. The protrusion from the suppressor cap is 0.26 mm at room temperature.

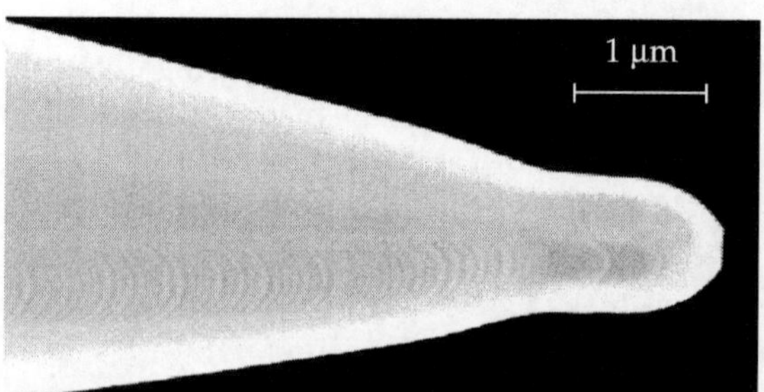

Figure 5.36 SEM image of the Zr/O/W{100} Schottky electron emitter before the experiments.

The emitter was mounted in a vacuum system with a background pressure of ~1 × 10^{-7} Pa. The evolution of the emission pattern has been determined using the setup described in Section 2.2 (Fig. 2.3, version A). Outside the vacuum the emission pattern was recorded through a view port with a CCD camera. The recording enabled us to detect both fast and slow changes: one ring series typically takes a few hours up to several days, which was detected with a time resolution of less than a second. To be able to detect the image on the screen, the beam leaving the extractor was accelerated to 7 kV before hitting the YAG. This gives a small focusing lens effect between the extractor and the screen, with the strength depending on the extraction voltage used. This effect has been calculated and was verified experimentally to be up to 8% for the recorded patterns. The results on pattern size have been corrected for this.

Forty-one consecutive ring series have been detected for a single tip at different operating conditions. Both temperature and extraction voltage have been varied, as shown in Fig. 5.37. The suppressor voltage was kept at –300 V with respect to the source at all times.

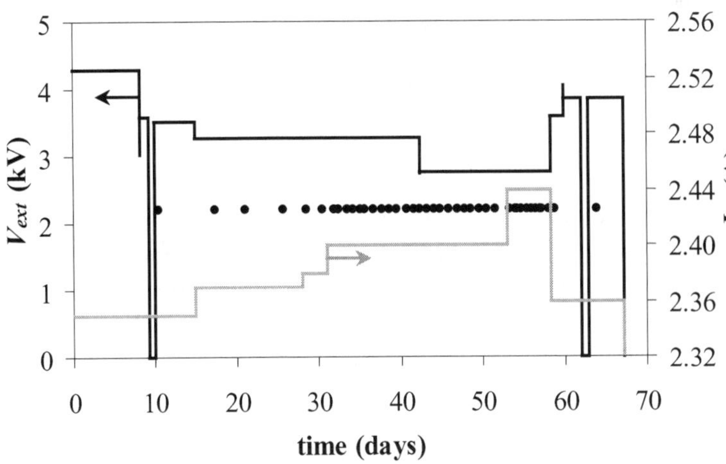

Figure 5.37 Operating conditions applied to the Schottky emitter in the experiment. Black dots indicate the end of a ring series.

The extraction voltage range was limited: the voltage had to be low enough to actually have collapsing rings, but for voltages too low, the number of emitted electrons became too low for an image on the

YAG screen. The temperature of the source was varied by changing the filament current.

For the typical range of operating conditions used in these collapse experiments, the concave current density distribution across the facet upon emission changes into a convex distribution on the screen (e.g., Fig. 3.13). As a result, the recorded emission pattern without rings is an approximately Gaussian intensity distribution. The Gaussian shape is the result of the spherical aberrations of the lens between the facet and the extractor: electrons launched further away from the facet center experience a stronger outward deflection than electron launched close to the facet center.

With respect to the interpretation of the pattern it is further noted again that the size of the emission pattern is not necessarily one-to-one with the facet size. For example, the beam divergence increases for increasing "sharpness" of the facet edge. This was illustrated in Fig. 3.26.

5.5.2.2 General system check: no-collapse operation

The emitter shown in Fig. 5.36 was stabilized by the manufacturer at an extraction voltage of 4.8 kV and a suppressor voltage of –240 V in the standard gun geometry. The SEM image of Fig. 5.36 is taken right after that stabilization. Prior to the experiments discussed in this chapter, the emitter was operated for a total of up to 48 hours but mostly at reduced temperatures. It is therefore not expected that the emitter size has changed considerably from that in Fig. 5.36.

Before inducing collapses, as a general system check, first the source was operated at an extraction voltage of 4.3 kV and a suppressor voltage of –300 V. These conditions were expected to give no collapsing rings. The field strength is calculated with the charge density method (http://www.electronoptics.com) for a rotationally symmetric emitter based on the contour of Fig. 5.36 in the standard gun geometry, taking into account 30 μm additional protrusion due to thermal expansion. This gives a field of 1.04 V/nm at the facet center. The minimum field for collapse prevention was calculated from Eq. (5.6) with the surface tension value found by Swanson and Schwind (2008). As a tip radius we take the radius of the inscribing circle that touches the end facet, which is 0.38 μm. This gives a minimum collapse prevention field of 0.96 V/nm.

The filament current was set at 2.35 A, which corresponds to 1,750 K, according to the manufacturer. The emission pattern did not display any rings in eight days' time (no marker in the first eight days in Fig. 5.36), and the facet current was ~22 μA. There was a gradual decrease in the facet current of ~10%, measured from a day after starting up the source. It might be that the extraction voltage was too low, and the current reduction is associated with a facet size reduction preluding a collapse (dictated by a global chemical potential gradient), or it could be the result of a restructuring in the tip area itself in an attempt to establish the equilibrium shape associated with these operating conditions.

After this system check the collapse experiment was started. The extraction voltage was reduced to 3.5 kV (and the facet current dropped to ~5 μA), yielding an estimated field strength at the facet of 0.80 V/nm. This is below the calculated minimum field strength to prevent collapse, and the operating voltage indeed caused a reduction of the pattern size and a drop of the total facet current of ~50% in 24 hours' time without any sign of stabilization. This indicates this voltage was too low and caused the emitter to start collapsing.

5.5.2.3 Collapsing analysis

In total, 41 ring series were initiated at different operating conditions. Figure 5.38 characterizes the series by different parameters: the full width half maximum (FWHM) of the emission pattern right after a series is finished, the current in that pattern, the ring series frequency, the "relative beam instability," and the symmetry of the ring series. The relative beam instability is defined as the ratio of the time in which the beam current displays the strong fluctuations and the total series time. This was determined from the intensity at the center of the pattern as a function of time. The symmetry of the ring series will addressed in the next section.

Changes in operating conditions do seem to affect some of the parameters in Fig. 5.38, as will be discussed below, but a third, perhaps more important, factor is also of influence, which is not controlled—the tip size. Figure 5.39 gives an SEM image of the emitter before and after the experiment.

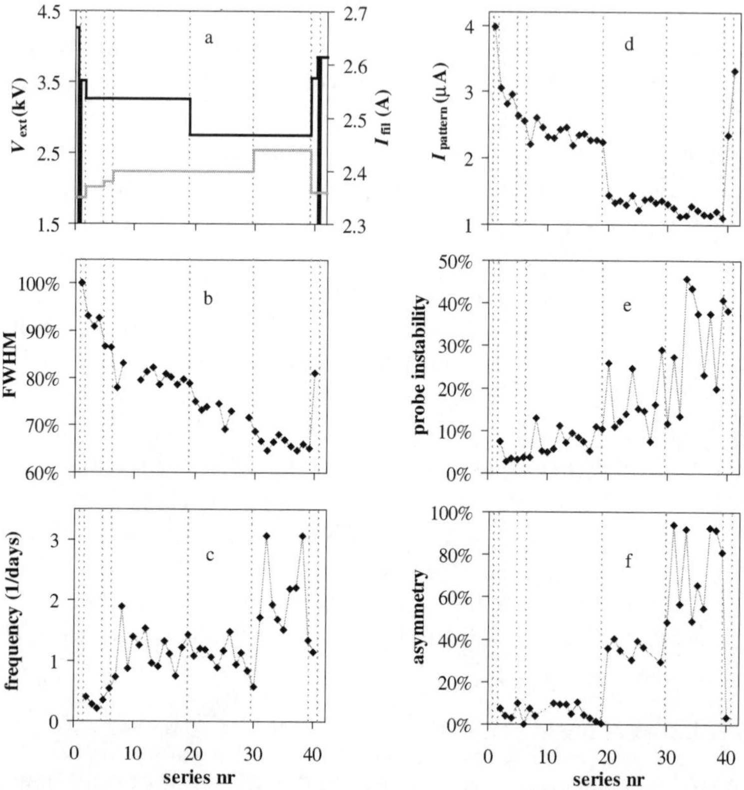

Figure 5.38 Different ring series characteristics as a function of the series number.

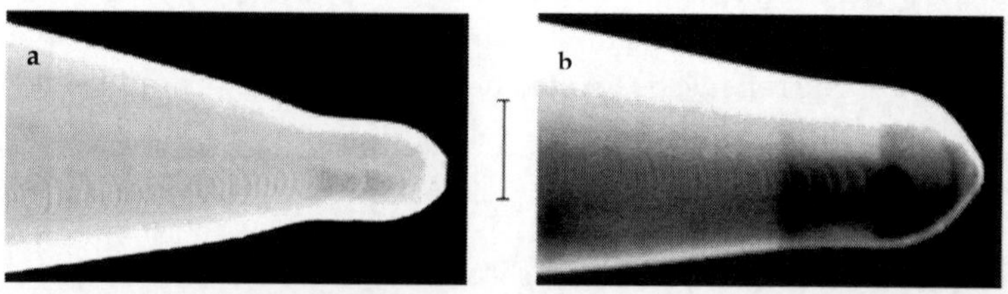

Figure 5.39 SEM image of a Zr/O/W(100) Schottky electron emitter before (a) and after (b) the experiments. The scale bar corresponds to 1 μm. Darkening is due to contamination induced by the imaging.

It can be seen that the tip size has increased considerably. Consequences of the tip growth are a reduction in the average field enhancement at the tip and a reduction in both the surface curvature and the field gradient between the tip and the shank. These changes are expected to affect the collapse behavior during the experiment, in addition to the changes in the operating temperature and voltage.

Up to the last data point in Fig. 5.38b, the FWHM of the pattern is showing a continuously decreasing trend, seemingly unaffected by the changes in operating conditions. We interpret this with a tip end becoming more spherical during the experiment in response to the relatively low operating voltage. This gives a smaller emission pattern, associated with a smaller end facet right after a collapse, despite the growing tip size. The effect of the changes in the operating conditions on the equilibrium shape of the tip end on the collapsing emitter is considered negligible with respect to this, and also the effect of the operating conditions and the geometry near the facet edge on the translation between pattern size and facet size is considered to be minor. The last FWHM in Fig. 5.38b is associated with the facet right after finishing a ring series at a voltage close to the voltage at the beginning of the experiment. It can be seen that the pattern size is much larger again, which is associated with a larger facet size, induced by the higher extraction voltage.

The ring series frequency (Fig. 5.38c) shows a trend that is generally opposite to that of the pattern size, although it does reflect the changes in temperature. The correlation with the pattern size might be understood by assuming the number of atoms involved per collapse decreases with decreasing pattern size. The correlation with temperature is considered the result of temperature-dependent diffusion rates. It is noted that the frequency response is quite different from that observed by Fransen *et al.* (1999), which confirms that the "history" of the emitter matters.

Perhaps surprisingly, the temperature does not seem to affect the current in the emission pattern (Fig. 5.38d). The pattern current *is* affected by the extraction voltage and also shows a correlation with the pattern size. The apparent temperature independence can (at least partly) be ascribed to the temperature dependence of the work function. Possibly the end facet becomes also slightly larger upon reducing the temperature as a result of a change in the equilibrium shape of the tip (Fig. 5.7). The dependence on voltage is ascribed to the field dependence of the effective work function (the Schottky effect).

The percentage of the operating time that the beam is strongly fluctuating shows an increasing trend (Fig. 5.38e), similar to that of the frequency and opposite to that of the pattern size. Assuming all other parameters are constant, the time that the beam is stable is indeed expected to decrease if the size of the facet to collapse is smaller, because the central area that is used for probe formation will be affected sooner.

Finally, several hours after series 40 was finished, the extraction voltage was increased to 3.8 kV. This voltage proved high enough to stop the collapsing, at least in the short term: the emission pattern was stable for about a week without any sign of collapse (except for collapse 41 [not fully monitored], which was induced by a hardware error that shut down the extractor voltage temporarily). The collapse prevention field is lower for larger tip sizes, Eq. (5.6), and the field associated with the extraction voltage of 3.8 kV applied to the tip in Fig. 5.38b is estimated to be ~0.6 V/nm, which is very close to the minimum field strength estimated to prevent collapse.

The experimental results illustrate that the tip-end geometry is a delicate interplay between effects on two different length scales.

5.5.3 (A)symmetry

In the previous section it was shown with Fig. 5.38f that the symmetry of the ring series can vary from 0% to 100%. The symmetry of the ring series is determined by the positioning of the island(s) on the facet with respect to the facet center, with, for a 0% symmetric ring series, a configuration with the islands at the facet edge. Figure 5.40 gives a few emission patterns and SEM images to illustrate different degrees of symmetry.

The data in Fig. 5.38f showed that the ring series were highly asymmetric, until the voltage was dropped, and became even more symmetric when the temperature was increased. And upon reducing the temperature and increasing the voltage, the ring series became highly asymmetric again (last data point in Fig. 5.38f).

The asymmetry might be related to anisotropy in properties such as surface tension, step line tension, and step stiffness, which can be enhanced upon applying a field (Ibach, 2006).

As an illustration of the effect of anisotropic properties, Fig. 5.41 gives a typical SEM image of an emitter that has been operated for many hours at a relatively high field. Note the different types of steps on the shanks.

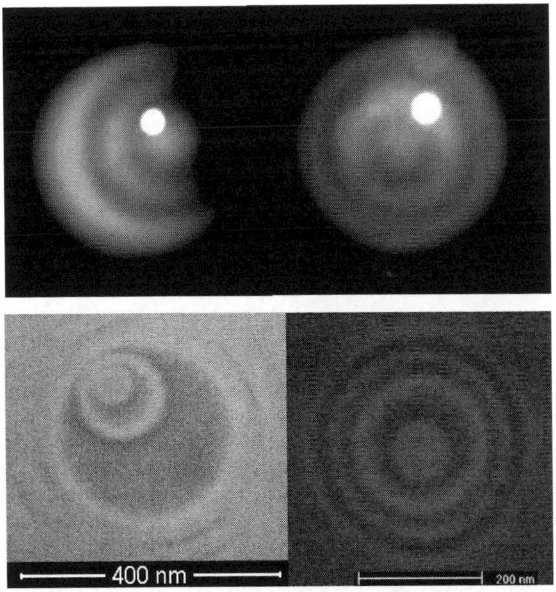

Figure 5.40 (Top row) Example of asymmetric and symmetric emission patterns in different ring series recorded. (Bottom) SEM images of two different emitters frozen in the middle of an asymmetric and a symmetric cycle.

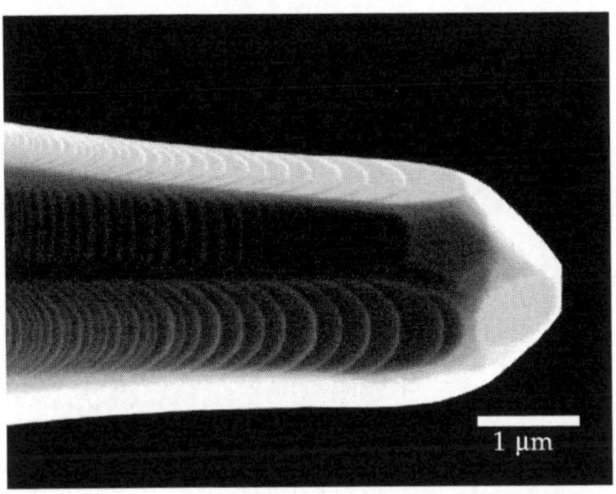

Figure 5.41 SEM image of a Schottky emitter showing steps on the shank. (Image from FEI Company.)

There is a difference in the steps for the different orientations: straight steps for the {110} planes and curved steps for the {100} planes. Note also that the terrace width is larger toward the tip where the cone angle becomes smaller. The step formation on the shanks is believed to be a form of thermodynamic faceting or phase separation (e.g., Liu *et al.*, 1997) that reduces the total surface energy. It is suggested that the straight staircases consist of alternating {110} and {112} microfacets, while the curved {100} terraces are separated by bunches of steps without a preferential step orientation. This "microfaceting" takes places everywhere on the emitter: Fig. 5.42 shows SEM images of stacks of {110} islands (left) and {100} islands (right) that can be found on the cylindrical part of emitters. Note that the {110} islands have both straight and curved sections, while the {100} islands are circular.

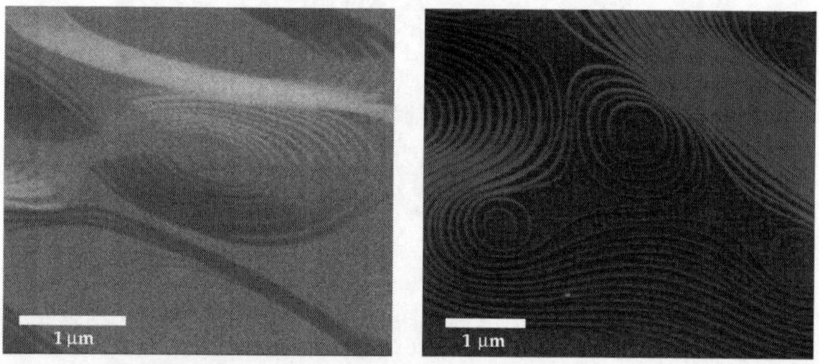

Figure 5.42 SEM images of {110} islands (left) and {100} islands (right) on the shank of Schottky emitters.

The {100} terraces on the cone and the cylindrical part of the emitter are basically no different from the collapsing {100} terraces at the tip end. Figure 5.43 shows a top-view and a side-view image of the same emitter tip. The tip end shows a circular end facet with two islands of different sizes stacked on top. Note the steps on the surface indicated by the black arrows. These are similar as those seen on the cone, but now it becomes clear that they develop with a preference for particular orientations. This may suggest a small difference in step stiffness for different step orientations. For illustration, in Fig. 5.44 ball models are shown of two different step orientations in the {100} plane.

The steps indicated in Fig. 5.43 are considered to be the "seeds" for the ring-shaped {100} terrace that forms in the collapse of the end facet and may be connected to the observed asymmetry of collapsing rings.

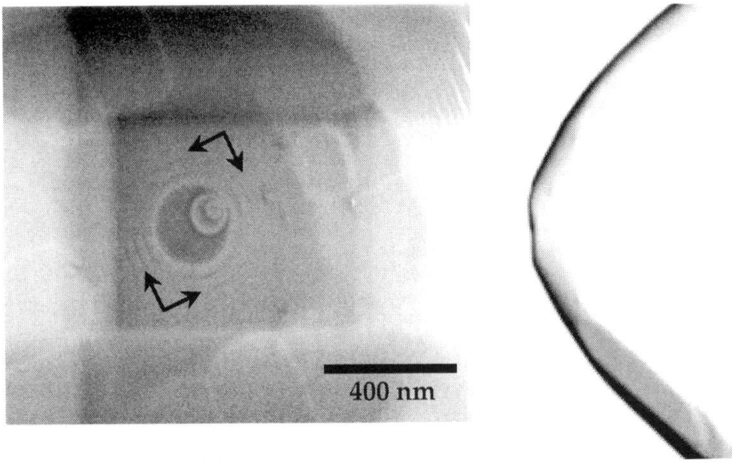

Figure 5.43 SEM images of a Schottky emitter, looking onto the tip end along the emitter axis (left) and from the side (right).

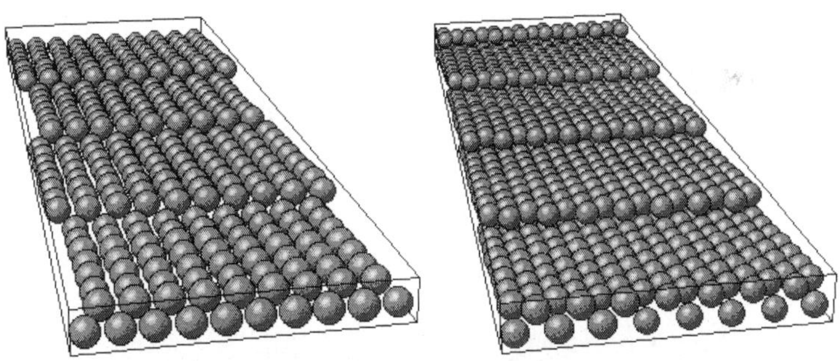

Figure 5.44 Ball models of a vicinal bcc(100) surface with steps along the (001) direction (left) and along the (011) direction (http://surfexp.fhi-berlin.mpg.de/).

But the asymmetric ring series from the experiments in the previous section (Section 5.5.2) do not seem to have a clear preference for a particular crystallographic orientation. This is concluded from Fig. 5.45. This graph is basically an overlay of the emission patterns all scaled to the same size. Each marker gives the position of the island in the pattern just before the ring series is completed (e.g., Fig. 5.46 at 14 h or Fig. 5.48b at 87.42 h). The dashed circle indicates the pattern edge.

Figure 5.45 Symmetry of the emission pattern is affected by the operating conditions. (Left) The position of the island remains on the facet with respect to the edge of the pattern (dashed circle) just before the series is finished for all ring series recorded. (Right) The emission pattern for the emitter showing the emission from the {100} terraces on the shanks.

The series recorded at low extraction voltage are in the center, while those recorded at high extraction voltage are on the edge. Figure 5.45 also shows the emission from the fourfold symmetry of the {100} terraces on the shank of the emitter to indicate the orientation of the emitter with respect to its axis.

Figure 5.46 shows the evolution of a ring series that was initiated at an operating voltage of 2.7 kV (and 2.44 A heating current) but finished at an increased operating voltage of 3.6 kV (and 2.36 A heating current). For operation at 2.7 kV the emitter was showing symmetrically collapsing rings with an average series time of ~10 hours, and the last cycle finished ~7 hours before increasing the voltage. The asymmetry of the pattern in the last few images of Fig. 5.46 does support the idea of a correlation between symmetry and the field strength at the tip end.

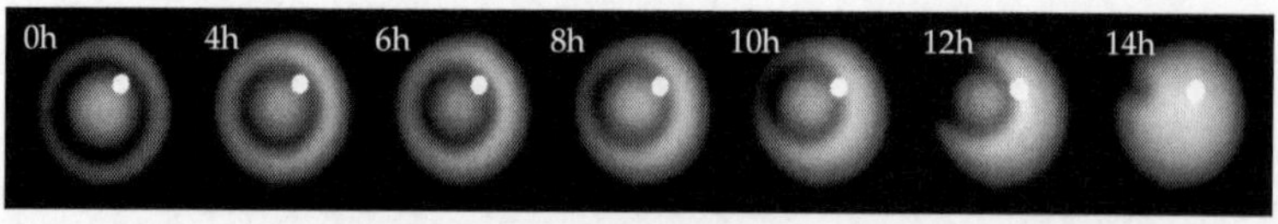

Figure 5.46 Evolution of the emission pattern, showing a change from symmetric to asymmetric. The patterns are recorded for an extraction voltage of 3.6 kV and a filament current of 2.36 A. The first pattern is taken right after applying these conditions.

At a higher extraction voltage the absolute field difference between the facet center and the facet edge is larger. The drift in

Fig. 5.46 suggests that from areas where the field is lower, on average more island atoms are detaching (and, if the island size remains the same, attach to the other side or, if the island shrinks, diffuse off the facet). Thermal fluctuations could initiate the drift.

Comparable asymmetry was also observed for a decaying mound on the Si(111)7×7 surface at 710–760 K (Ichimiya *et al.*, 2001), for the room-temperature decay of double-layer islands on Cu(111) (Giesen *et al.*, 1998), and for multilayer islands on Au{100} in 0.05 M H_2SO_4 aqueous solution at 1.05 V (Hirai *et al.*, 2001). In these studies the imaging techniques used are scanning tunneling microscopy (STM) and electrochemical atomic force microscopy (EC-AFM), and although the island diameters are of approximately similar size, the height of the structures is only a couple of atomic layers. The number of atoms involved in these studies is thus at least smaller by a factor of 10^4. The asymmetry is not completely understood.

5.5.4 Detailed Geometrical Description

As shown in the first section of this chapter, the collapsing rings on Schottky emitters disturb the beam current of systems. To clarify the origin of the typical profile of Fig. 5.5 we first need to know what exactly happens during a ring series. This is investigated below.

We will use the evolution of the emission pattern and the associated "total tip-end current" profile for ring series number 3 from Fig. 5.37 or 5.38, recorded at 2.37 A (~1,775 K) and an extraction voltage of 3.3 kV. These conditions are closest to operation for applications that require a low-energy spread. The particular ring series is not the first one recorded, so the exact tip geometry at the onset of this series is unknown. If we take the geometry of Fig. 5.36 the calculated field strength is ~0.75 V/nm. For a work function of 3.03 eV this would be associated with an intrinsic practical brightness of $\sim 1 \times 10^7$ A/m^{2}srV and an intrinsic full width containing 50% (FW50) energy spread of <0.25 eV. It is noted that the energy spread of the beam at the target is likely to be larger due to Coulomb interactions between electrons in the beam on the path from source to target.

Figure 5.47a gives the total tip-end current during ring series number 3, from 0 hours to 100 hours. Figure 5.48a gives a few of the associated emission patterns. Because we are interested in the typical beam current fluctuations, we also "reconstruct" the beam

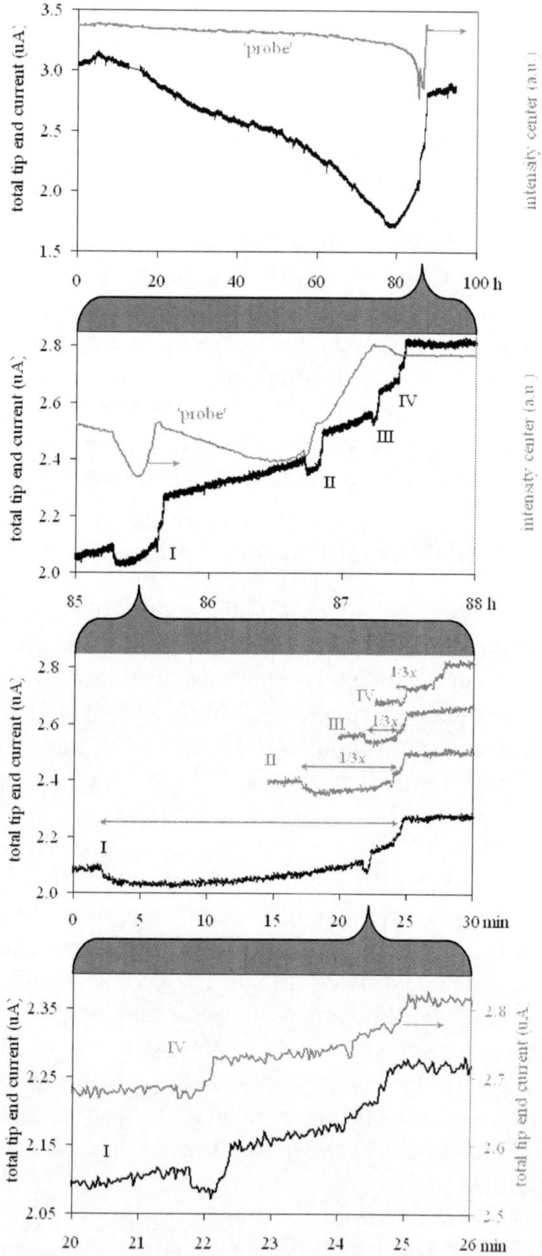

Figure 5.47 Total tip-end current for ring series number 3 (Figs. 5.37 and 5.38), recorded at I_{fil} = 2.37 A and a 3.3 kV extraction voltage.

by taking the central intensity of the recorded emission patterns. The variation in this intensity is plotted in gray with the total tip-end current in Fig. 5.47a. There is a good resemblance with the beam current profile in Fig 5.5.

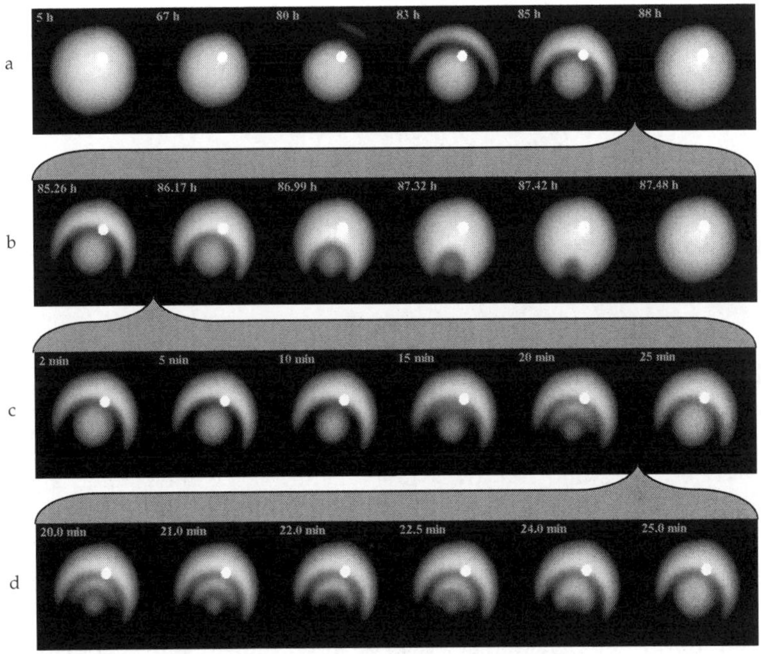

Figure 5.48 Emission patterns associated with ring series number 3 (Figs. 5.37 and 5.38), recorded at I_{fil} = 2.37 A and a 3.3 kV extraction voltage.

In Fig. 5.47a it can be seen that in the first 80 hours the total tip-end current drops gradually to 60% of its original value. Figure 5.48a shows that in this period the emission pattern decreases in diameter. We interpret this as a facet decreasing in size.

After 80 hours the current in Fig. 5.47a starts to increase again, and in the emission pattern a bright ring appears (pattern 83 h in Fig. 5.48a). We associate this pattern with a tip-end geometry similar to that shown in the SEM image of Fig. 5.49.

It is believed that a large step has developed on the curved surface surrounding the facet, changing the {100} end facet into a {100} island on top of a larger {100} facet. The height of the island is typically a few tens of nanometers. The slopes of the island have a higher work function, and the area on the lower facet close to the island is shielded from the electric field. This can explain the dark ring in the emission pattern.

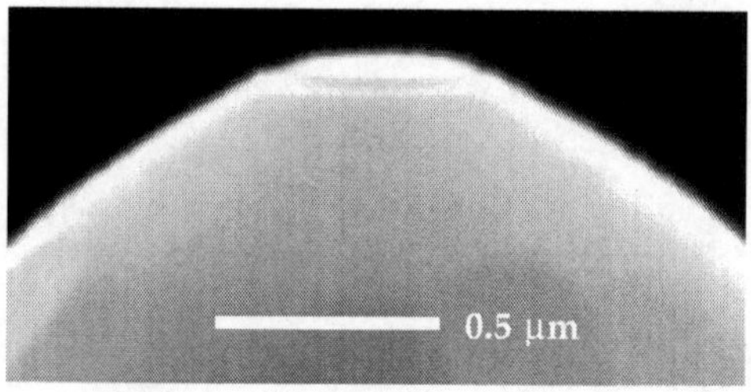

Figure 5.49 SEM image of the typical one-step geometry of a Schottky emitter during a ring collapse.

By comparing the emission patterns for 83 hours and 85 hours (Fig. 5.48a) it can be seen that the diameter of the dark ring decreases. We associate this with a decreasing island diameter. With decreasing island size the low emission area surrounding the island also decreases and the low work function facet underneath gets more exposed. This explains why the total tip-end current increases.

Between 85 hours and 88 hours the total tip-end current in Fig. 5.47a rapidly increases to a value close to its original value. The emission pattern at 88 hours closely resembles the original pattern. Our interpretation is that between 85 hours and 88 hours the island atoms have all been transported away to the shank and that the tip end has a single facet again, similar to the starting geometry.

The reconstructed beam current profile in Fig. 5.47a resembles that of Fig. 5.5 and shows that contrary to the total tip-end current, the beam is relatively undisturbed by the geometry changes, except for the last few hours of the profile. The description above cannot explain the double dip that is observed. We need to zoom in on the section between 85 hours and 88 hours. Figure 5.47b gives the total tip-end current and reconstructed beam in this time frame, while Fig. 5.48b gives a few of the associated emission patterns. First we will describe the geometry changes, and we will return to the beam in Section 5.6.2.

The emission patterns in Fig. 5.48b show a decreasing diameter of the dark ring. Accordingly, the total tip-end current in Fig. 5.47b shows an increasing trend. This is consistent with the interpretation of a decreasing island diameter. The curve in Fig. 5.47b is, however, not a smooth function. This is an indication there is more going on than simple island diameter reduction. We have marked four sections: I–IV. They look rather similar, although increasingly less pronounced, suggesting some kind of repetitive process might be involved. We zoom in on the changes in Section I in the 86th hour, in Figs. 5.47c and 5.48c.

At two minutes, in Fig. 5.47c, the current shows a sudden drop. This drop is not reflected by a change in the emission pattern other than an overall intensity reduction. We interpret this event with the formation of a step on the slope of the {100} island, which "splits" the island into two islands lying on top of each other. The SEM images in Fig. 5.50 give a few examples of such a two-island geometry in a more developed state: in these images the diameter of the top island is already considerably smaller than the island underneath.

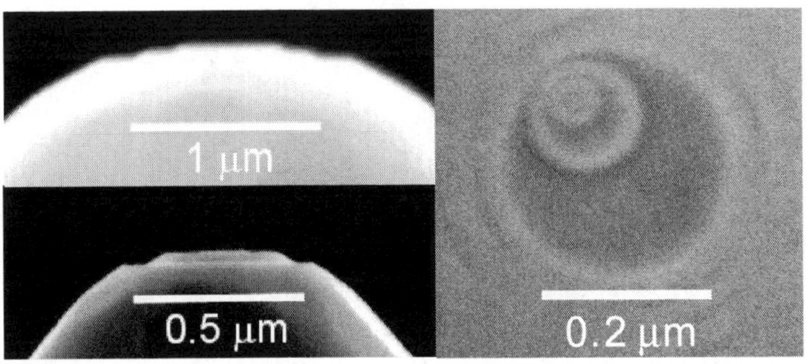

Figure 5.50 SEM images of the two islands' geometry on Schottky emitters.

A sudden reconstruction on the slopes of the island as proposed can change the field distribution on the tip end. This could explain the sudden drop in intensity. Simulations show that the average field strength across the tip is smaller for a geometry with a smooth island slope, compared to that with a slope with a step, and would give reduced emission. At the same time, the newly created {100} ring on the slope is not developed enough to contribute to the emission yet.

After the drop at two minutes, the current in Fig. 5.47c decreases a little bit more until it reaches a minimum, around eight minutes. Then the current starts to increase again. In the emission patterns in Fig. 5.48c it can be seen that first the dark ring expands inward, and then a bright ring appears within the dark ring. We associate these changes with first a reduction in size of the top island, exposing more of the underlying island. Then, when the new ring-shaped {100} area has developed enough, it starts contributing to the current and it appears as a bright ring in the emission pattern.

The current in Fig. 5.47c increases progressively from 8 minutes onward and comes to an abrupt halt at 25 minutes. We notice the irregularities in the current between 20 minutes and 25 minutes, but we will come back to these later. The emission pattern after 25 minutes (Fig. 5.48c) shows that the smallest of the dark rings has disappeared. We interpret this with the relocation of all atoms of the top island to somewhere on the emitter shank and full exposure of the island underneath. The abrupt trend change in the current in Fig. 5.47c at 25 minutes is thus associated with the completion of the clean-up of the top-island atoms.

Summarizing, we interpret the changes in Section I as a split-up of the original island into two islands on top of each other, after which the atoms of the top island are all transported elsewhere, and the tip end is left with a single island again. The net effect of Section I is thus a reduction of the island height.

It is noted that the current level after Section I is considerably higher than before Section I, while the emission patterns before and after (first and last emission pattern in Fig. 5.48c) show a similar geometry. By looking closely, however, it can be seen that the dark ring in the pattern after Section I (25 min) is thinner. We ascribe this to the reduced island height. An island of smaller height will have a smaller slope area and will yield less shielding of the facet underneath. This can explain why in Figs. 5.47b and 5.47c, despite a similar tip-end geometry, the current right after Section I is larger than just before.

Figure 5.47c also gives the current profiles for Sections II, III, and IV, which are shifted back on the time axis. The shape of those current profiles looks similar to that in Section I, and we therefore ascribe them all to geometry changes involving a reduction of the height of the island.

Summarizing, our interpretation of Fig. 5.47b (85–88 h) is that the island on the facet reduces in height in Sections I, II, III, and IV and decreases in diameter in between these sections. Because both island height and diameter become smaller, the current variations in Sections II, III, and IV are increasingly faster and less pronounced.

Figure 5.47d compares the last five minutes of the ring series, just before it is finished (gray curve), with the last five minutes of Section I. It can be seen that the current profiles are in close agreement. The low resolution of the emission pattern and the small intensity variations make the interpretation difficult, but it is clear that in the last five minutes of the ring series the remains of the original island that developed at 80 hours must be cleaned up. These last five minutes include Section IV, indicated in Fig. 5.47d, for which we concluded that the island height reduces, and it must be that in the second half of that curve in Fig. 5.47d the last atoms are removed. Because of the similarity of the current profiles we associate the same sequence of events with the current variations in the last five minutes of Section I (in black in Fig. 5.47d). The difference is the geometry of the tip end beneath the disappearing island: in the last five minutes of the ring series the disappearing island lies directly on the facet, while for Section I the disappearing island lies on top of another island present on the facet. Figure 5.48d gives a few of the associated emission patterns for Section I. The interpretation of these is that the top island undergoes a height reduction between 21.0 minutes and 22.5 minutes. The SEM image in Fig. 5.50b, which shows an emitter with an island of rather small height on top of a much thicker island on the facet, fits with this description.

In Fig. 5.47d the current variations are very small, but it seems that beyond 23 minutes, there might be more height reduction steps involved before the island atoms are all transported away. Looking back at the current profiles in Section II and III in Fig. 5.47c, the dips in the second half of these profiles resemble those in Fig. 5.47d, and these may also be interpreted with the same sequence of events: island splitting and dissolution of the upper island.

Altogether we see at least four times that the base island on the facet splits into two islands on top of each other (in the first half of I, II, and III and in IV). And for most of these times we have seen that at least once the top island splits again (in the second half of I, II, and III). The ring series is finished when all atoms of the originally formed island have been transported elsewhere, and the single-facet geometry is restored.

The correlation between these changes and the probe current will be addressed in the next section, where the effect of geometry changes on the beam properties is discussed.

5.6 The Effect on Beam Properties*

The changes in tip geometry affect two parameters relevant to the performance of microscopes and lithography machines, the field enhancement at the facet center and the lens effect between the central facet area and the extractor. Together, these effects will change the probe current–probe size relation or the relation between probe current and beam coherence.

The general effect of tip size growth at constant operating parameters is a reduction in brightness and energy spread, as the field enhancement at the tip drops, and an increase in virtual source size, as the divergence of the central beamlet decreases (Fig. 3.30). Changes in facet size can cause variations on the general trends connected to tip size growth.

5.6.1 Facet Size Changes

It has been shown above that by changing the operating conditions, the facet size can increase or decrease (e.g., Figs. 5.17, 5.24, and 5.47a), and it can stabilize (e.g., Fig. 5.21) or start to collapse.

For a reduction in facet size, the field enhancement at the facet center increases. This increases the emitted current density and thus the brightness and energy spread of the beam. Secondly, a reduction in facet size changes the lens effect between the facet and the extractor: the beamlet from the facet center becomes more diverging. This decreases the virtual source size.

The first effect enhances the beam current (or angular intensity), but the second effect suppresses it. The net effect on the beam current depends on the exact conditions. Figure 5.51 gives two examples of the expected changes, for a reduction of the facet size for an emitter operated at 0.8 V/nm on the left and for an increase in facet size for a larger emitter operated at a field of 1.0 V/nm on the right.

In both cases, the beam current is relatively constant and conceals that important properties such as brightness, energy spread, and virtual source are actually changing considerably. Translated to practical systems, for a shrinking end facet size, the beam current may be constant, but the geometrical image in scanning transmission electron microscopy (STEM) decreases, while the chromatic aberration increases, or for transmission electron

*Most of this paragraph has been previously published and was reproduced with permission from *J. Vac. Sci. Technol. B*, **27**, pp. 2524–2030, Copyright (2009), and reprinted from *Ultramicroscopy*, **110**, pp. 1243–1254, Copyright (2010), with permission from Elsevier (see also page 249 of this book).

microscopy (TEM), the spatial coherence increases but the temporal coherence decreases.

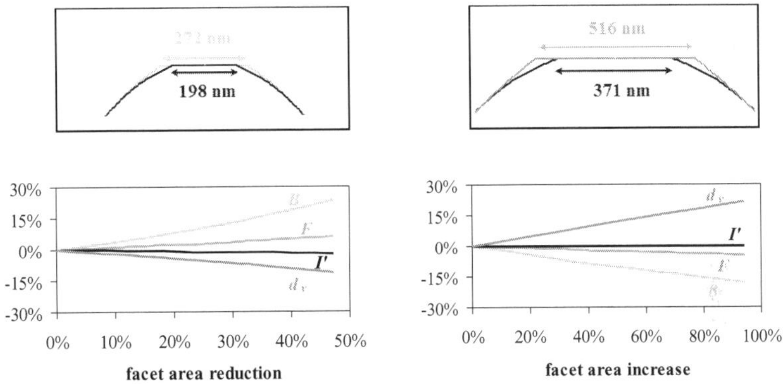

Figure 5.51 Change in field strength, brightness, virtual source size, and angular intensity (beam current) for a reduction of the facet area (left) and an increase of the facet area (right). Calculations are performed for the standard gun geometry with a suppressor voltage of −0.3 kV and an extraction voltage of 3.3 kV (left) and 6.0 kV (right), yielding an initial field strength at the facet center of 0.8 V/nm and 1.0 V/nm, respectively.

Depending on the application (at high beam energies generally chromatic aberration is negligible) the facet size adjustments can gradually increase or decrease the resolution of the system until the end facet size stabilizes or collapses (and temporarily destroys the beam properties completely).

The performance of a system is sometimes changed deliberately by changing the extraction voltage. This will instantly change the brightness, energy spread, and probe current due to the change in field strength, but the preceding sections have shown that in the following hours, up to days, a more gradual change should also be expected, which either undoes the initial change a little bit or further enhances it, as a result of the response in the facet geometry. Figure 5.51 shows that these changes may not be noticed by monitoring the probe current.

Note that in Fig. 5.21, in the first few hours upon increasing the voltage from 2 kV to 6 kV, there actually is a change in beam current (the central pattern intensity). This trend is, however, also seen in both the screen current and the extractor current and might be ascribed to an effect different from a change in facet size. After the first few hours the changes in beam current are small, while the screen current is clearly still increasing. This confirms that the probe current is not necessarily a good indicator of changes of the tip geometry that induce changes in brightness and energy spread.

5.6.2 Facet Collapse

With the detailed description of the geometry changes that take place in a ring series from Section 5.5.4 and the simulations in the previous paragraph, we can now interpret the typical beam current profiles of Figs. 5.5 and 5.47a.

It was shown that in the first phase of the collapse the facet diameter decreases, but the probe current in Figs. 5.5 and 5.47a remains approximately constant. This can now be understood from the simulations in Fig. 5.51: the current density at the facet increases, but at the same time the beam also gets spread out more by the facet extractor lens. The beam current really starts dropping when the island slope moves into the area that delivers the beam current. Figure 5.47b gives a close-up of the reconstructed dips between 85 hours and 88 hours. The first dip is associated with the first height reduction of the island on the facet (Section I). In this phase a relatively large portion of the tip end displays low emission, which is likely to affect the area delivering the beam electrons. The second dip is broader and less deep and is associated with the removal of the remaining island atoms. Because for Sections II through IV the island is smaller than in Section I, and resides off center near the facet edge, the beam current, which in a properly aligned system originates from the central area, does not reflect all the geometry changes in great detail.

The reconstructed beam current profile of Fig. 5.47a,b is a typical beam current profile for a collapsing Schottky emitter, but in practice no two beam current profiles are exactly alike. This can be explained by the fact that the properties of the area delivering the beam electrons are never exactly the same: the geometry of the tip end in a collapse is usually asymmetric, and the position of the island(s) on the facet is random (Fig. 5.46). To illustrate the effect of this on the profile, Fig. 5.52 gives three reconstructed beam current profiles for the same ring series we have discussed in detail above, where for each profile a different area on the facet has been taken.

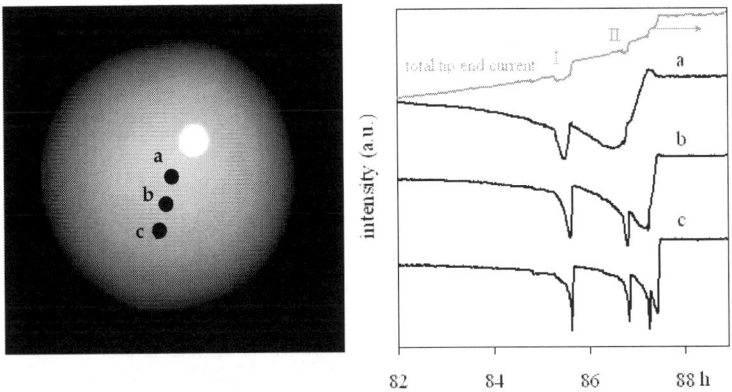

Figure 5.52 Three beam current profiles during a ring series, with different offsets, reproduced from local intensities at different positions in the same emission pattern series.

Profile **c** in Fig. 5.52 gives the intensity variation for the position on the facet that is the center of the islands (see, e.g., the 20 min pattern in Fig. 5.48c). For this position multiple dips are observed. The exact nature of the beam current profile will thus be different for different alignments of the emitter with respect to the optical axis of the gun and will also vary if the position of the islands on the facet is not always the same. On the basis of Fig. 5.52 we can say that for properly aligned systems, profile **a**, with two dips with the second dip much broader, is typical for asymmetric ring series (relatively high fields), while profile **c**, with multiple dips, is typical for very symmetric ring series (low fields).

5.7 Concluding Remarks

The main conclusion of this chapter is that the Schottky emitter is "alive," so to speak. It does not have a static shape. The shape dynamics are an interplay of the thermodynamic equilibrium shape of the tip end and that of the emitter as a whole. As a result, in practice the tip size of standard Schottky emitters grows with increasing operation time.

Generally, at low operating fields the tip size grows with material that peels off at the tip end (collapses); at higher operating fields the tip size grows with material from the shank.

An increasing tip size means a decreasing field enhancement at the tip and a decreasing divergence of the beamlet from the facet center. The field enhancement and lens effect are, however, also affected by the actual size of the end facet, which can give variations on that main trend line.

The facet size is affected by the local field strength. Whether the facet grows or shrinks upon a change of the voltage depends on the initial degree of faceting of the tip and the effect of the change on the equilibrium crystal shape of the tip and the direction of the chemical potential gradient in the tip vicinity. Although the changes in facet size cause considerable variations in the brightness, energy spread, and virtual source size of the beam from the facet center, this is generally not reflected by any change in the beam current for applications, because the effect of the change in the field and the beam divergence cancel out.

With respect to stability it is relevant that polyhedral tip geometries have a relatively stable end facet. This is ascribed to the energy barrier involved in nucleating a new layer on the facet. Such energy barriers can prevent geometric changes if the change in the conditions are not large enough (e.g., the "threshold" voltage found by Fujita and Shimoyama [2007] that marked the sudden disappearance of the {211} facets on a polyhedral W(100) emitter tip) or the facets too large (e.g., the comment by Swanson and Schwind [2008] that the {211} facets on large Schottky emitter tips did not get "extinguished" as on smaller emitters).

Different tip sizes may have slightly different field ranges in which they can be operated without becoming unstable (collapse or become unstable with respect to microprotrusions). That is, smaller tips require a higher field strength to prevent collapsing (lower field limit) but may also have a higher upper field limit with respect to geometrical stability: for the same field at the facet center, smaller emitters will have a relatively lower field at the ridges and corners separating the facets.

Finally, note that for two emitters with different tip sizes but with an equivalent field strength at the tip, the equilibrium crystal shape of the tip is expected to be the same (except for extreme cases) but the chemical potential gradient between the tip and the shank will not. This is because the surface curvature on the cone away from the tip area will be no different, but the field at the cone is stronger for the larger emitter due to the lower field enhancement at the tip, which requires a higher extraction voltage (or lower suppressor voltage) to set the same tip field. Similarly, for two equivalent tips with a different cone angle or cone shape, the tip shape is expected to be same if operated at an equivalent field at the tip but the chemical potential gradient between the tip and the shank will be different. This is likely to affect tip growth as a function of time. Possibly there is with respect to stability an optimum combination of tip size, cone shape, and field strength.

In the next chapter we will combine the results of this and the previous chapter in search for the maximum performance for different applications and the associated optimum operating conditions.

Chapter 6

Optimum Operation

The preceding chapters have provided the necessary ingredients to finally address the question about the limits of Schottky source performance.

It has been shown in Chapter 4 that the performance of a system in terms of probe size–probe current, or probe current and coherence, is limited by the brightness and energy spread of the beam and the aberration coefficients of the column. And it has also been shown that the brightness and energy spread of the beam are interrelated. Which combination of brightness and energy spread will give the best performance depends on the specific application. And which operating conditions are associated with that optimum performance depends on the emitter geometry. Chapter 5 has shown that the possible range of operating conditions is limited by the geometrical stability of the tip. Section 6.1 brings this together and discusses the maximum possible performance for different applications.

To be able to set the right operating conditions that translate to the desired optimal brightness–energy spread combination, it is required that it be known how the applied voltages and heating current translate to a temperature and field strength for the center of the end facet of the emitter. The voltage–field relation is a strong function of the emitter geometry (Chapter 3), and in Chapter 5 it has been shown that the emitter shape evolves in operation. To maintain optimum emission properties one should thus adjust the operating

Physics of Schottky Electron Sources: Theory and Optimum Operation
Merijn Bronsgeest

ISBN 978-981-4364-79-9 (Hardcover), 978-981-4364-80-5 (eBook)
www.panstanford.com

conditions during the emitter lifetime. Section 6.2 addresses possible source-monitoring tools to keep track of the changes. Finally, Section 6.3 gives some practical considerations.

6.1 Maximum Performance for Different Applications

The optimum combination of brightness and energy spread is not the same for every application. This is because for different applications different probe size contributions are dominant (Section 4.3).

For high-resolution work at low beam energies usually chromatic aberration and diffraction contributions dominate the probe. If one can keep the contribution from the source image size negligible, and work at beam currents below $1 \times 10^{-18} B$ A (Kruit & Jansen, 2008), it can be shown that performance can be increased by reducing the energy spread. The appropriate figure of merit is in that case thus $1/\Delta E$. There are two limitations, but these will be addressed later.

For larger desired beam currents the contributions from chromatic aberration and the geometric source image dominate the probe, and it can be shown that performance can be increased by increasing the ratio $B/\Delta E^2$. The above would apply to, for example, scanning electron microscopy (SEM) or C_s-corrected scanning transmission electron microscopy (STEM).

For systems in which chromatic aberration plays no role of importance, such as STEM, the figure of merit is the brightness B.

And a final example is systems with a monochromator: in that case the figure of merit can be shown to be approximately $B/\Delta E$.

6.1.1 Maximum Performance from a Static Emitter Shape

Below we will assume the emitter shape is static and discuss the optimum choice in emitter size and operating conditions for each of these applications. We compare the four emitters from Fig. 3.27 in the standard gun geometry. The beam properties are calculated on the basis of the emission theory in Chapter 1, and we use the strategy of Section 4.4.1 to take into account the effect of Coulomb interactions.

When $1/\Delta E$ is the figure of merit, performance can be increased by reducing the energy spread but, as said above, not indefinitely. A first limitation is related to the fact that by reducing the energy spread the brightness also decreases (see Fig. 4.30). A lower brightness means that, not to lose in beam current, the source image has to become larger. This narrows down the probe current range for which the source image is negligible with respect to the contribution from diffraction ($I_p < 1 \times 10^{-18} B$ A), and $1/\Delta E$ still is the appropriate figure of merit. The limit is where the minimum required beam current becomes equal to $1 \times 10^{-18} B$ A. Beyond this point the performance will start to decrease if the energy spread is further reduced. With respect to this it can be seen in Fig. 4.30 that at equivalent brightness B, the smaller emitter has the smallest energy spread, and it thus seems preferable.

Figure 6.1 gives results for different figures of merit, as a function of the operating conditions, for a beam-defining aperture at 1 mm from the emitting surface. The work function used for the calculations is given by Eq. (3.16). Figure 6.2 gives the results as a function of the angular current density, which is the parameter that can be verified experimentally relatively easy.

We would like to point out that the results apply to the particular parameter set and calculation methods, as discussed above, and need to be recalculated for, for example, a different value of the work function or a different position of the evaluation plane.

Figure 6.1 shows that the energy spread can be reduced by reducing the extraction voltage, but with respect to the other figures of merit, Figs. 6.1 and 6.2 show that B, $B/\Delta E$, and $B/\Delta E^2$ display a maximum. Note that for each emitter the operating voltage and angular intensity for maximum performance is different for the different figures of merit. The optimum operating conditions of an emitter are thus dependent on the application of interest. The figures also show that the maximum brightness value is larger for larger sources, while the maximum $B/\Delta E$ and $B/\Delta E^2$ are larger for smaller sources. This suggests that smaller sources should be preferred over larger sources such as SEM, C_s-corrected STEM, or systems with monochromators, if the desired probe current is larger than $1 \times 10^{-6} B$ pA. The results also predict that at lower temperature performance will be slightly better.

For evaluation planes further away from the emitting surface, the maximum achievable performance decreases due to a larger

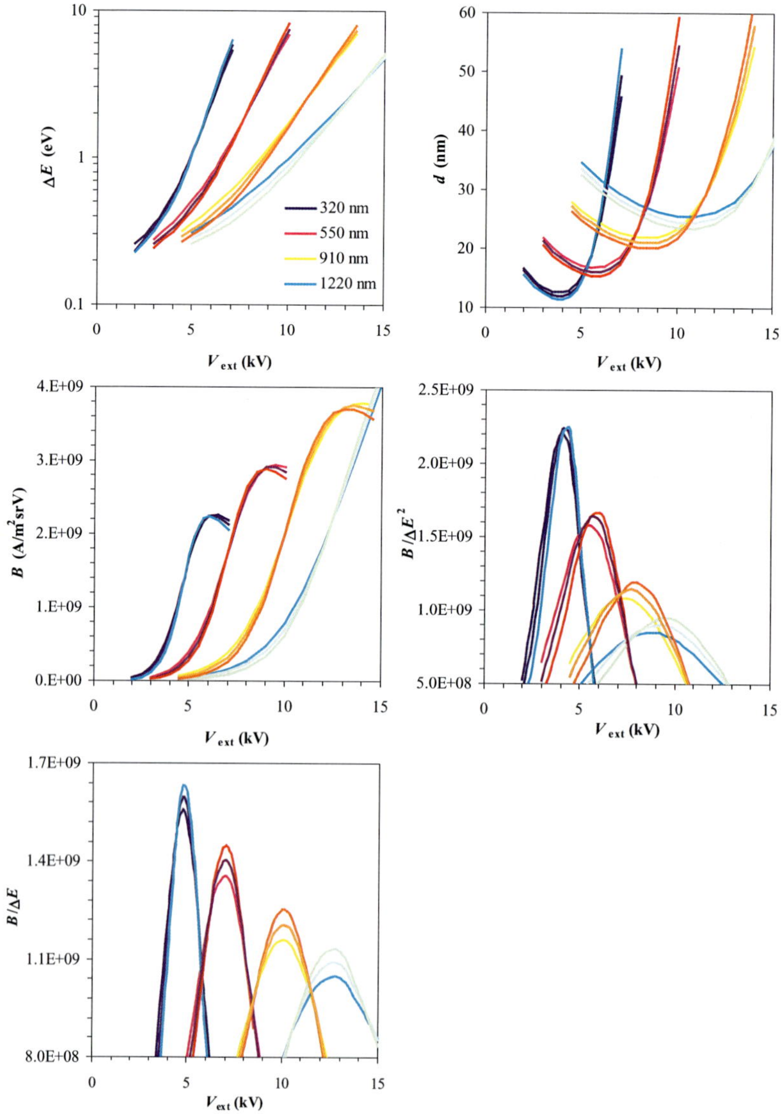

Figure 6.1 FW50 total energy spread, FW50 virtual source size, B, $B/\Delta E^2$, and $B/\Delta E$ as a function of the extraction voltage for 1,600 K, 1,700 K, and 1,800 K and for four different emitters (Fig. 3.27). Evaluated at 1 mm from the emitting surface. *Abbreviation*: FW50, full width containing 50%.

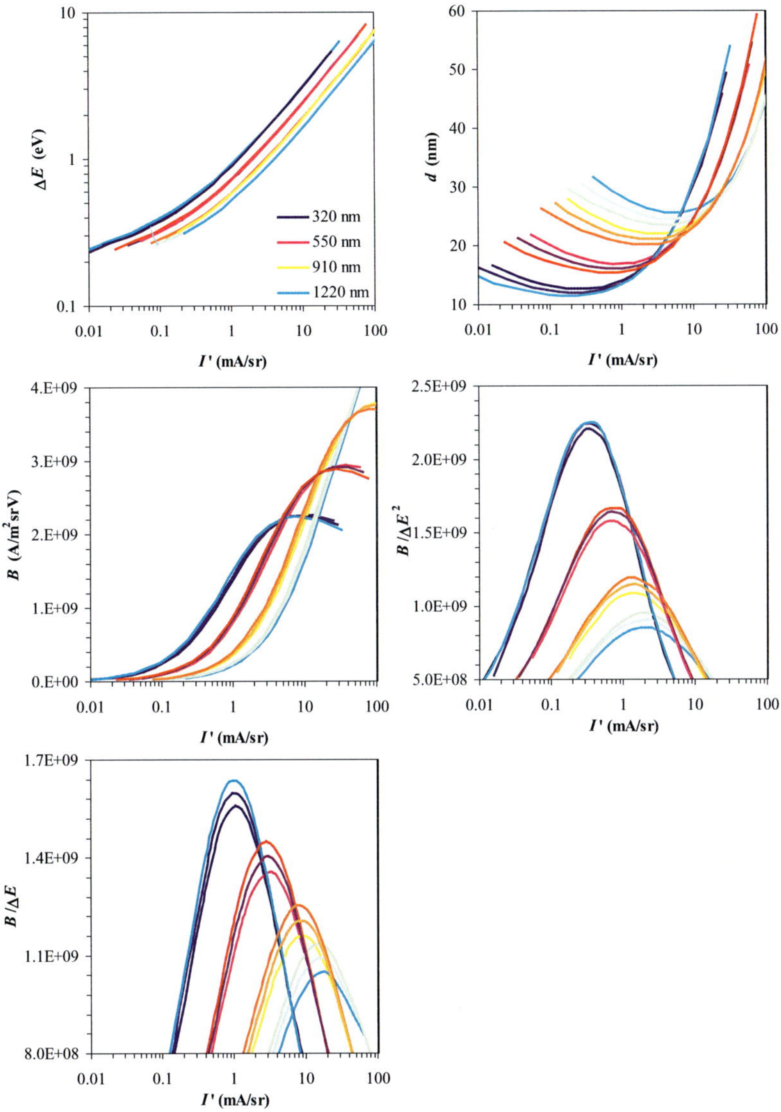

Figure 6.2 FW50 total energy spread, FW50 virtual source size, B, $B/\Delta E^2$, and $B/\Delta E$ as a function of the angular intensity for 1,600 K, 1,700 K, and 1,800 K and for four different emitters (Fig. 3.27). Evaluated at 1 mm from the emitting surface.

brightness loss (e.g., Fig. 4.21). This is shown in Fig. 6.3. Aperturing the beam, as first recommended by Jiang (1996), can prevent such additional loss. Note that the maximum achievable performance at each z-position in Fig. 6.3 is associated with slightly different settings of the extraction voltage.

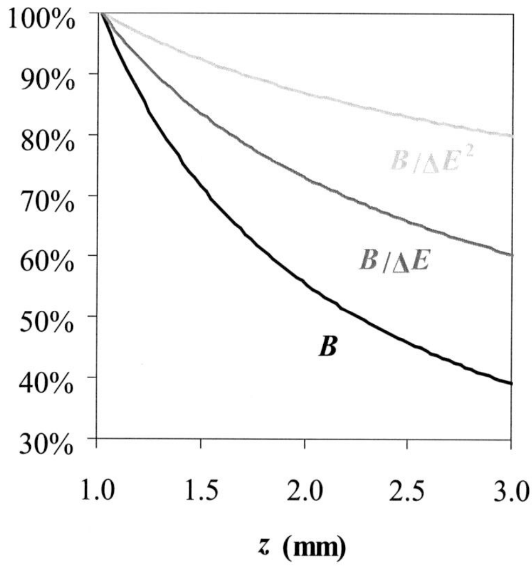

Figure 6.3 Beam length dependence of the maximum achievable values of B, $B/\Delta E$, and $B/\Delta E^2$ for the 550 nm emitter at 1,800 K. The extraction voltage and angular intensity associated with these optima decrease with increasing beam length.

6.1.2 Geometrical Limitations

We have shown that for applications limited by chromatic aberration and diffraction performance can be increased by reducing the energy spread. It was also shown that reducing the energy spread also reduces the brightness and thus increases the relative contribution of the geometrical source image. Above it was concluded that in this respect a smaller emitter was preferable because it has a smaller energy spread at equivalent brightness. However, a second limitation to reducing the energy spread is the risk of geometrical instability of the end facet of the emitter: for operation at too low fields (too low energy spread and brightness) the end facet will start collapsing. According to Eq. (5.6) the minimum field to prevent collapsing is proportional to $1/\sqrt{r_{tip}}$ and thus larger for smaller emitters. For the smallest and largest tip geometry in Fig. 3.27, with tip "radii" of 0.32 μm and 1.22 μm, the estimated energy spread and brightness at 1 mm of the emitting surface, for operation at their respective threshold fields and at a temperature of 1,800 K, are 0.34 eV and 1.8×10^8 A/m^{2}srV against 0.26 eV and 1.4×10^7 A/m^{2}srV. If one can work at very low beam currents (lower than $1 \times 10^{-18} \times 1.4 \times 10^7 = 14$ pA), and the facet instability is to be avoided, the larger emitter would thus the better choice. The estimated minimum extraction voltage and angular intensity to prevent instability of the facets are indicated in the first images of Figs. 6.1 and 6.2 by the black textured line. Note that the calculated no-collapse angular intensity is considerably lower than the experimental value reported by Swanson and Schwind (2008). Possible explanations are a different definition of tip radius, a different tip end geometry, and/or a different work function for the emitters used here in comparison to the emitter for which the minimum no-collapse field was derived.

It is noted that because the no-collapse field is higher for smaller emitters (Sections 5.3.1 and 5.6.1), the shape of the end facet associated with the no-collapse field might be less circular in comparison to that for larger emitters (different equilibrium crystal shape).

The maximum brightness in Fig. 6.1 is obtained for very high extraction voltages, which give a high field strength at the emitter surface and for the large emitters also a relatively high field strength between the suppressor and the extractor. For the given parameter set, the maximum is found for a field strength of ~2.3 V/nm and angular intensities >7 mA/sr. The question is if such conditions can give stable emission.

There is no record of an experimental realization of such conditions. According to Schwind *et al.* (2004), a conventional Schottky emitter in a conventional gun geometry operates at a maximum angular intensity of 1.0 mA/sr. The maximum experimental angular intensity, reported in the literature, for stable emission (at least 70 h) from a Schottky emitter at 1,800 K with a reported radius of 0.8 μm, and in an only slightly different gun geometry, is 2.8 mA/sr, which was associated with a field strength of 1.2 V/nm (Van Veen, 2000). As pointed out in preceding chapters an accurate translation from operating conditions (heating current, voltages) and measurement variables, such as angular current density, to the field strength and temperature at the emitting surface is not that straightforward, and the error in the reported field value may thus be considerable. The geometry of the emitter associated with the high angular intensity and field was verified with SEM after the experiment: the tip was fully faceted, with the {100} end facet connected to four {110} facets.

Van Veen (2000) has explored the limits and managed to get 4.4 mA/sr and 6.3 mA/sr, but for those conditions the emission was unstable. The pressure increased considerably, and the power supply ran into its current limit of 1 mA. It is noted that for operation at such high fields (>1.2 V/nm) the extractor material has to deal with the large shank currents (heating, electron-stimulated desorption (ESD), etc.), and enhanced emission from sharp edges together with increased gas pressure due to ESD could induce arcing, which could destroy the emitter. As discussed in Section 5.2, for very high field strengths the surface becomes undefined and can start to develop microprotrusions. All of the above could explain the instability of the emission. For equivalent field strength at the facet center, larger emitters have a considerably higher field strength at the facet edge than smaller emitters. It might be that because of this, smaller emitters can give stable emission at slightly higher field strengths at the facet center.

The maximum field strength that can still give stable emission has not been investigated yet, but the result by Van Veen (2000) suggests that a 0.8 μm radius Schottky emitter may not be stable geometrically for field strengths much larger than 1.2 V/nm at the facet center. This sheds a different light on the results in Figs. 6.1 and 6.2. The field strength that is associated with a maximum $B/\Delta E$ ratio is ~1.7 V/nm and for $B/\Delta E^2$ ~1.4 V/nm. It is noted that the calculations apply to a specific tip geometry, and the field at the facet can deviate by roughly 10% for different sizes of the end facet, but this suggests that for most applications optimum operation is associated with a fully faceted tip end showing only the end facet and the {110} facets; that is, assuming the same gun geometry and work function of the surface and a beam-limiting aperture at 1 mm from the emitting surface. For these parameters the maximum practical brightness is thus not limited by the effect of Coulomb interactions but by the geometrical stability of the tip and is roughly ~5×10^8 A/m^{2}srV if we take 1.3 V/nm as the upper field limit. The energy spread associated with this maximum is affected by the Boersch effect, yielding an FW50 of ~0.5 eV for the smallest emitter, increasing to >0.8 eV for the largest emitter in these simulations.

Figure 6.4 gives the results as a function of field strength at the facet center with a rough indication of the relevant field range for geometrical stability. The inset in the first figure is a rough illustration of the equilibrium shape of the end facet for different field strengths. It is noted that due to these geometry changes the properties of the beam will also change. The given curves, for a constant shape, are thus only approximates.

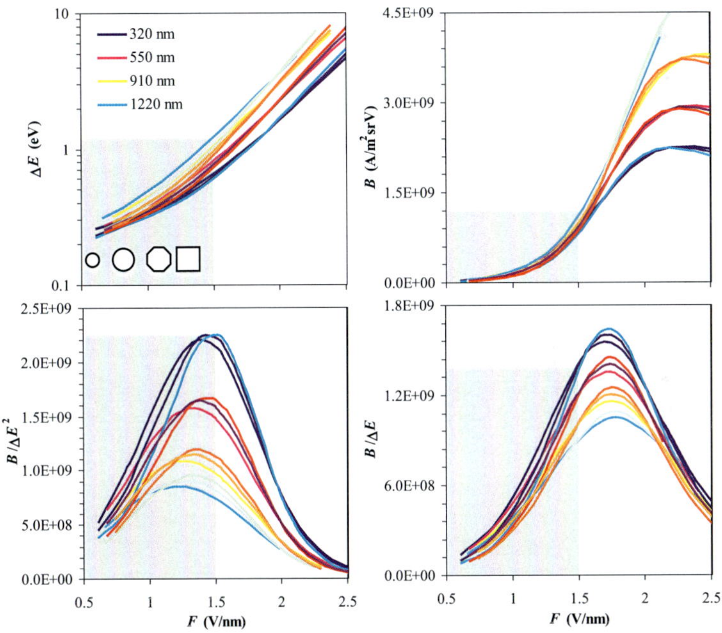

Figure 6.4 Data of Figs. 6.1 and 6.2 as a function of field strength. The gray boxes indicate approximately the field range for which the tip can be stable geometrically (no collapses, no microprotrusions). The inset in the first figure gives the trend in the equilibrium shape of the end facet expected as a function of field strength.

If the beam is limited further down the column the optimum in B, $B/\Delta E$, and $B/\Delta E^2$ shifts to lower fields (maximum possible performance drops, see Fig. 6.3). For example, for a beam-limiting aperture at 10 mm from the emitter we calculate an optimum field of ~1.1 V/nm for $B/\Delta E^2$ (optimum field for $B/\Delta E$ and B is still very high

[~1.3 V/nmand ~1.9 V/nm]), which is close to the field for which the emitter in Fig. 5.12 was repeatedly stabilized to show an octagonal end facet and both the {110} and {211} facets on the tip. In this case the maximum performance for, for example, SEM or C_s-corrected STEM (figure of merit $B/\Delta E^2$), is limited by Coulomb interactions, while for other applications (figure of merit B or $B/\Delta E$) the maximum performance is still limited by the geometrical stability. Note that the tip geometry associated with optimum operation would be different for the different applications: an octagonal end facet for applications limited by $B/\Delta E^2$ and a square end facet for applications for which $B/\Delta E$ or B is important.

6.2 Source-Monitoring Tools

Considering the observed geometry changes and their effects on beam properties it is important to monitor the source if one is interested in getting the most from the system. Monitoring allows one to keep track of changes of the emitter tip geometry, to prevent the end facet from collapsing, and to tune the operating conditions for optimum performance.

It is noted that not only the tip geometry might change but also the relation between heating current and tip temperature might change due to, for example, wire thinning by evaporation. And the work function might change due to contamination or due to depletion of the ZrO_x reservoir. It is therefore recommended to monitor as many parameters as possible.

Useful monitoring tools are, for example, the Schottky plot, the total emission current, and the facet emission pattern. Unfortunately it is not always possible to measure these parameters. Below we will give a few examples.

It is noted that a parameter that would be useful to monitor, but that is usually not available, would be the facet current, or the difference between the total emission current and the current collected at the extractor (for extractor apertures that accept the full facet beam). Monitoring the facet current would be useful, for example, because it would allow one to detect a collapse in an early stage, when the beam fluctuations associated with a collapse can still be avoided by increasing the extraction voltage, before the island-on-facet geometry has developed.

Finally, it is noted that combining the results of different monitor tools can be useful, for example, to distinguish changes in facet size from changes in temperature or work function.

6.2.1 Schottky Plot Slope

Section 5.4.2 has already shown how the Schottky plot slope can track changes of the end facet. Those data were taken on a test system. Appendix A gives procedures to record Schottky plots for a few commercial systems, which is slightly less straightforward. Just as an example we give a Schottky plot measured for a Quanta 200F scanning electron microscope in Fig. 6.5.

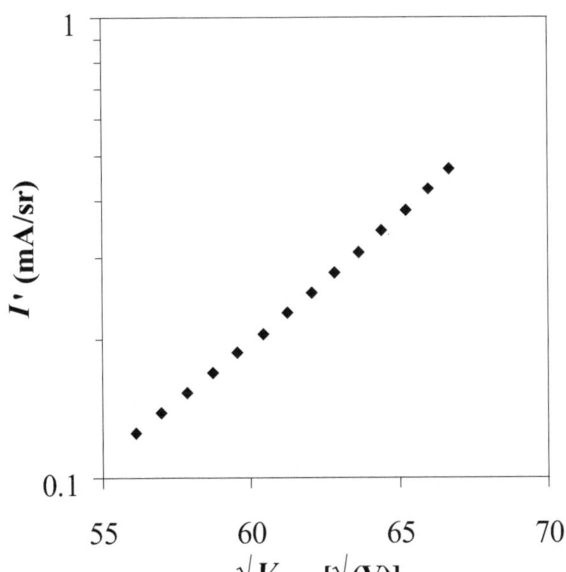

Figure 6.5 Example of a Schottky plot recorded for a commercial system (Quanta 200F scanning electron microscope). The procedure is given in appendix A.

The trend in the Schottky plot slope as a function of time can be used to estimate the tip growth and changes of the end facet geometry, as was shown in Fig. 5.28. It is recommended to take Schottky plot data regularly.

6.2.2 Total Emission Current

Figure 6.6 shows the changes in the total emission current as a function of time that we measured for constant operating conditions for a Schottky emitter in a commercial system. It can be seen that the total emission increases with more than 50%. This is ascribed to geometry changes of the emitter, which yield more {100} surface area. It was verified with SEM that the tip had increased in size considerably during its lifetime.

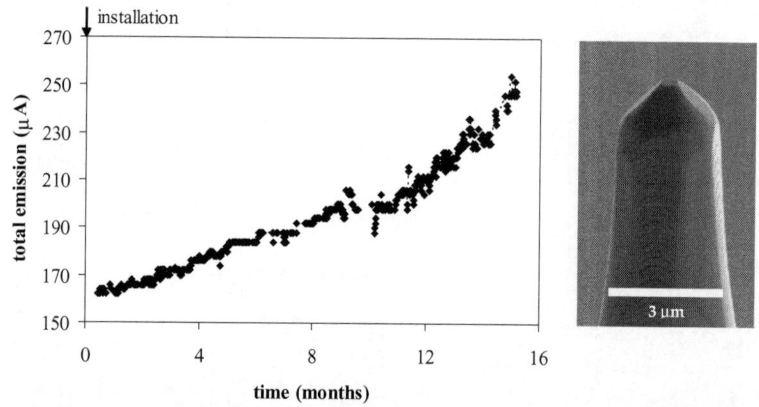

Figure 6.6 Total emission current as a function of time for a Schottky emitter in a commercial system (Quanta 200F) at constant suppressor and extraction voltage and heating current. The SEM image is taken after the emitter was taken out, which was after 21 months. Total emission data for months 15–21 have not been included, because they contain a lot of noise, related to a loose contact that was accidently introduced during a service inspection, but the trend was not disturbed.

6.2.3 Facet Emission Pattern

Some systems allow us to measure the shape of part of the facet emission pattern. Appendix A describes how this can be done for the Quanta 200F scanning electron microscope. The shape of the pattern (whether it is concave, flat, or convex or shows rings) can give useful information about the situation at the facet (see Chapter 3). Figure 6.7 gives an example of a measurement of the shape of the facet emission pattern for two different extraction voltages, showing that it changes from convex to concave. For this system the facet emission pattern can be probed only partially.

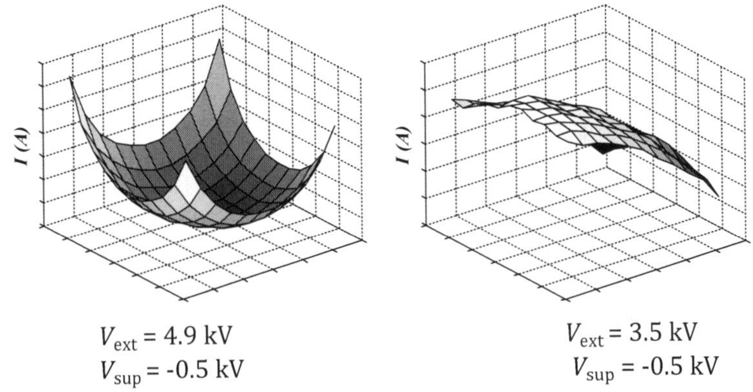

Figure 6.7 Part of the facet emission pattern recorded in a scanning electron microscope (Quanta 200F) for two different extraction voltages. Procedure in appendix A.

6.3 Practical Considerations

6.3.1 For Users

New Schottky emitters are generally stabilized to a polyhedral shape. It is recommended for a new source to operate it at exactly the same conditions as the emitter was stabilized at by the source manufacturer. Measure the angular intensity, total emission, and the Schottky plot slope and verify these are equivalent to the values from the manufacturer.

Monitor what you can in the first few hours up to days to check on geometry changes. And keep monitoring the source regularly during the lifetime of the emitter and make adjustments to the voltage settings for maximum performance, if required. Preferably, do not increase the emitter temperature: the effect on the emission properties is limited, and it speeds up the irreversible tip size growth. To change the field there are different options, but note that by changing the suppressor voltage the energy of the beam leaving the extractor and entering the gun lens can be kept constant.

After a change in voltage the system may require several hours to stabilize, because the change in voltage can cause changes of the tip end geometry. Keep track of this by temporarily increasing the monitoring frequency.

Geometry changes will not only change the brightness and energy spread of the beam but also change the virtual source size. Adjusting the voltage settings can bring back the original brightness and energy spread, but these settings are not likely to give the original virtual source size as the latter is a function of the geometry-dependent lens effect. For maximum performance, adjustments of the optics might thus also be required during the emitter life.

6.3.2 For System Manufacturers and Experimental Setups

In general, for all systems it is recommended that before a new source is installed, it be investigated, which is the optimum operating field for the particular application in terms of desired emission properties. This requires knowledge of the source size, the gun geometry, and the position of the beam-limiting aperture. Then check if this field could give a geometrically stable tip (field not too high, field not too low; range depends on tip size but as a rough estimate 1.0–1.5 V/nm for a tip radius of ~0.3 μm up to 0.5–1.2 V/nm for a tip radius of ~1.0 μm) and decide on the preferred field.

In the production process, new Schottky emitters are generally stabilized to a polyhedral shape. Upon installation of a new source it is recommended to initially operate it at exactly the same conditions as the emitter was stabilized at by the source manufacturer. Measure the angular intensity, total emission, and the Schottky plot slope and verify these are equivalent to the values from the manufacturer.

Then change to the desired field by adjusting the voltage settings. Keep in mind that the surface tension changes with the square of field, so the effect of a change in extraction voltage on the equilibrium tip shape is expected to be larger for larger voltages. If the desired field is not much below the field for which the tip was stabilized, it might be wise to first decrease the field considerably, to remove atoms from the tip area while it is reshaping to a more spherical geometry, and then to set the new conditions, instead of going straight to the new conditions and risk that the tip end becomes more spherical using atoms that are brought in by the shank and thus grows in size immediately.

Monitor the angular intensity, total emission, and Schottky plot slope regularly in the first few hours to check on the geometry changes. Start calibrating the optics only after the emitter geometry and with that the virtual source size of the beam have stabilized.

Keep monitoring the source during the lifetime of the emitter. If possible, monitor the brightness and energy spread of the beam. Adjust the operating parameters and optics, if required.

It is noted that for special applications the generally undesirable geometry changes could also be exploited, for example, to temporarily increase the brightness of the probe at a given voltage: operate the source for several hours at a low voltage to reduce the facet size and increase the field enhancement at the facet center and then set back the voltage. The brightness of the probe will be temporarily increased and will degrade again in the next few hours. A similar strategy could serve to temporarily decrease the energy spread.

If a system is not used permanently one could consider to change to different operating conditions for idle time, or shut down the source completely if the system is not going to be used for a long time, to avoid unnecessary tip size growth.

A sure option to reduce the rate of geometry changes without turning off the source completely is to reduce the operating temperature. This will reduce the diffusion and evaporation rates. For the standard emitter module, changing temperature can change the alignment of the emitter with respect to the suppressor aperture and the extractor aperture due to asymmetry in the construction (see, e.g., Fig. 2.10). Changing the temperature is thus not recommended for systems with stringent alignment requirements, unless the design of the emitter module prevents changes of the alignment for changes in temperature.

A potentially interesting option comes from the response of an operating Schottky emitter that ran into the current limitation of the power supply two hours after the extraction voltage had been increased: although the tip end was not polyhedral yet, the evolution of the emission pattern stopped upon reaching the current limit and continued only upon reducing the temperature (causing the emission to drop below the current limit) and upon increasing the current limit of the power supply. Shape changes that would increase the emission were apparently opposed in the current limited mode.

Finally, it should be possible to minimize changes of the geometry by choosing the right voltage settings. Currently, however, we do not have enough information to be sure about the right path to take. Staying near the no-collapse field (with a more round facet geometry) ensures at least temporarily zero or low net diffusion in the tip area, but the chemical potential at the tip is still higher than at the shank. Keeping the tip in a polyhedral shape has the advantage of the energy barriers that need to be overcome to realize shape changes, but the chemical potential at the tip is lower than that at the shank. Perhaps the best option is to keep the tip polyhedral at the lowest possible field strength.

Over the last 30 years the Schottky electron source has proven to be a reliable workhorse with excellent performance. And the above shows that by paying a little bit more attention to it, the performance can be easily increased.

For very high-end applications, the beam properties of the Schottky source may suffice, but its geometrical instability will become intolerable, and it is expected that the success of this source type can be sustained only if that can be tackled.

Appendix A

Procedures for Monitoring in a Few Commercial Systems

The main monitoring tool addressed here is the Schottky plot. At the end we will briefly address the recording of (part of) the facet emission pattern in the Quanta 200F scanning electron microscope (SEM).

Key to useful Schottky plot data from commercial systems are creating a field-free zone between the extractor and the beam-defining aperture (or at least zero lens effect), and making the beam path free from obstructions from the beam-defining aperture to the entrance aperture of a Faraday cup on the stage.

The apertures that can be used as beam-defining apertures for Schottky plot recording are for the:

NOVA NanoSEM: the 2nd extractor aperture (acceptance half-angle 3.45 mrad)

Quanta 3D FEG: the aperture after the C0 lens (acceptance half-angle 1.99 mrad)

Quanta 200F: the decelerating lens aperture (acceptance half-angle 1.34 mrad)

For the Quanta 200F the zone between the extractor and the beam-defining aperture can be made field free by turning off the C1 and C2 lenses. The objective lens is then used to focus the beam into the Faraday cup aperture.

For the Quanta 3D FEG the zone between the extractor and the beam-defining aperture can be made field free by turning off the C0 lens. For both the NOVA NanoSEM and the Quanta 3D FEG, to ensure a free beam path beyond the beam-defining aperture, the C1 and C2 lenses are used to make crossovers, with the C2 lens crossover onto the decelerating lens aperture. Note that when the extraction voltage is changed, the lens effect between the extractor and the C1 lens is also changed. It may be required, for each data point, to manually optimize the settings of the condenser system (C1 C2) to get all current through the decelerating lens aperture.

A.1 Recording Schottky Plots in Quanta 3D FEG and NOVA NanoSEM Systems

- Log on as "support" so that you are allowed to change the voltage settings of the electron gun.
- Vent the chamber.
- Put a Faraday cup on the stage (e.g., the one on a standard FEI* test sample).
- Free the beam path:
    - Remove the "insert" from the objective lens, if present, and leave the "insert" inside chamber (check the manual; use the designated tool).
    - Change to the largest aperture on the strip (1 mm diameter).
- Connect the Faraday cup to a current meter.
- Pump down the chamber.
- Bring the stage to a working distance.
- Change the configuration in "8_system configuration" to high vacuum mode.
- Quanta 3D FEG: go to spot 4.0, 10 kV beam energy; NOVA NanoSEM: go to spot 6.0 (the condenser systems of these machines are calibrated to give a focus on the decelerating lens aperture for these settings).
- Switch on the beam.
- Optimize the crossover position.
- Focus on the aperture of the Faraday cup in "quad 1" (top-left window).
- Go to the TstEGun window. Check the "advance view" box, if not checked.

---

*FEI Beam Technology Division, 5350 NE Dawson Creek Dr., Hillsboro, OR 97124, USA.

- Write down any relevant parameter values (date, system, heating current, heating voltage, gun pressure, total emission current, Coulomb tube (CT) current, extraction voltage, suppressor voltage, C1 lens setting, etc.).
- Turn off the beam.
- Change the operating mode from OPERATE to SERVICE. (N.B. The software Software can return a message, which you can "hide").
- Check the C1 lens voltage. Set it at the same value as it was in the OPERATE mode, if it has changed.
- Turn the beam on by pressing "HV on" in the TstEGun window (either of the two buttons).
- Change the ramping speeds for suppressor and extractor voltage from 0 V to 10 V for both up and down. Press "set speeds" to confirm.
- Go from scan mode to spot mode and put the spot inside the aperture.
- Write down V_{ext} ("Actual" value), V_{sup} ("Actual" value), I_{total} (FEG emission), I_{cup} (current meter), and the time in a table.

  1. Change from spot mode to scan mode.
  2. Reduce V_{ext} with 100 V.
  3. Change V_{sup} proportional to the V_{ext} change. This is to maintain the same field form around the tip (e.g., going from a 5.0 kV and 500 V to a 4.9 kV extraction voltage requires a suppressor voltage of 4.9/5.0 × 500 V).
  4. Go to spot mode and put the beam inside the cup (optional: check that the crossover position is still optimal).
  5. Write down V_{ext}, V_{sup}, I_{total}, I_{cup}, and the time in the table.

- Repeat the above five steps for the desired number of voltage settings. Check at one of the lower extraction voltages if the crossover is still looking OK and if all of the beam still goes into the Faraday cup.
- Bring back the settings to the settings of OPERATE mode.
- Beam off.
- Bring the extractor and the suppressor voltage back to the settings of OPERATE mode (see the top part of your filled out source monitoring work sheet). **Wait until the values are set.**

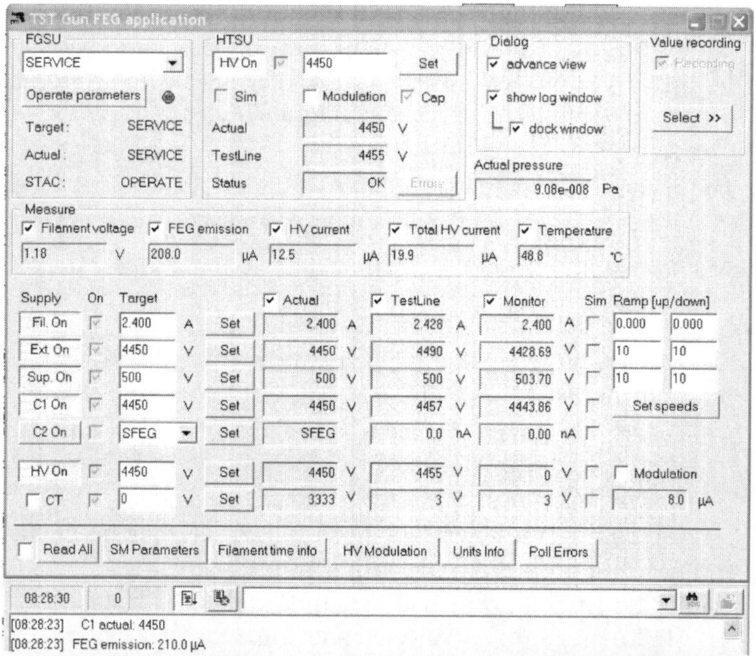

- Change ramping speeds of the extraction and the suppressor voltage from 10 to 0 (do not forget to press "set speeds").
- Change from SERVICE mode to OPERATE mode.
- Set back the aperture number from 1 mm to what it was before the measurement.
- Beam on and check the system by imaging something at high resolution.
- Beam off.
- Vent the chamber.
- Take out the Faraday cup.

- Put the "insert" back into the objective lens.
- Check Part III on the source monitoring sheet.
- Pump the chamber.
- Set the system configuration back from "high vacuum mode" to the setting that belongs to the "insert."
- Log off.
- Analyze the data and extract the field enhancement factor. Has it changed with respect to the last time, and what is the field strength the source is operating at?

Recording Schottky plots in the Quanta 200F system:

- Log on as "support" to be able to change the voltage settings.
- Vent the chamber.
- Put a Faraday cup on the stage (e.g., the standard FEI test sample).
- Free the beam path:
    - Remove the "insert" from the objective lens, if present (see the manual), and leave the "insert" inside the chamber.
    - Check if the blanker is present? Set it to the largest separation: setting 4.
    - Change to the largest aperture on the strip (number 1, 1 mm diameter).
- Connect the Faraday cup to a current meter.
- Pump down the chamber.
- Bring the stage to a working distance.
- Switch on the beam.
- Set the source tilt to zero.
- Focus on the aperture of the Faraday cup in "quad 1" (top-left window).
- Go to the TstEGun window. Check the "advance view" box.
- Write down any relevant parameter values (date, system, heating current, heating voltage, total emission current, CT current, extraction voltage, suppressor voltage, etc.).
- Turn off the beam.
- Change the operating mode to SERVICE. (N.B. The software returns a message, which you can "hide").
- Turn the beam on by pressing "HV on" in the TstEGun window (either of the two buttons).

- Change the ramping speeds for the suppressor and the extractor voltage to 10 V for both up and down. Press "set speeds."
- Turn off the CT (clear the box) (means it goes to ground potential).
- Change V_{C1} in the "Target" column to the V_{ext} value (press the set button or use Enter to confirm) (N.B. By doing this the potential on the C1 lens becomes zero with respect to ground).
- Change HV in the "Target" column to the V_{ext} value (press the set button or use Enter to confirm) (N.B. By doing this the potential on the extractor becomes zero with respect to ground).
- Go to the TstEOptics window. Open the xT_ObjectiveLensOutput window (objective/final lens) and the xT_CondensorLensOutput window (C2 lens).

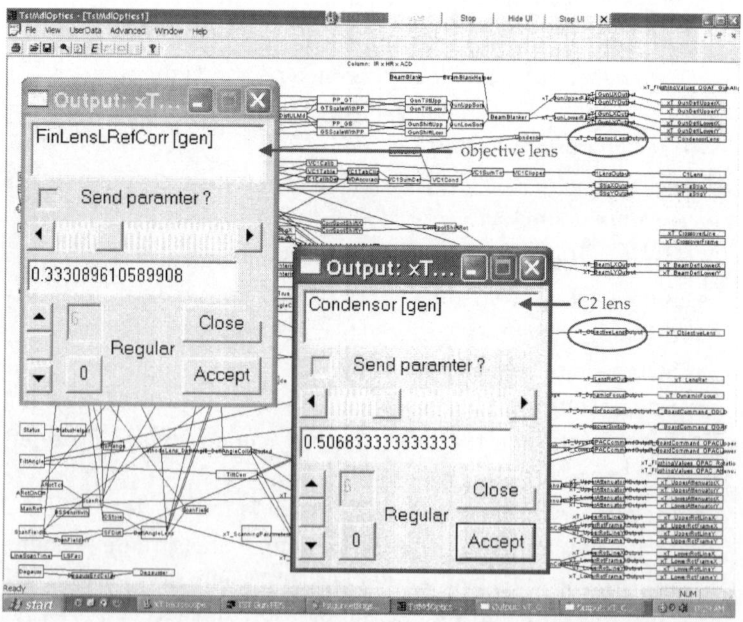

- Set the condenser lens excitation (C2 lens) to zero. Watch the image of the Faraday cup aperture and adjust the objective lens setting if you start to lose track of the hole.
- If necessary focus a little bit by optimizing the objective lens setting.

- Go from scan mode (full frame) to spot mode and put the spot inside the aperture.
- Check the current meter reading to make sure all current is going into the cup.
- Check if the alignment of the aperture on the stick is such that all current passes it: do this by changing its alignment and monitor the current reading from the cup and maximize it.
  - Write down V_{ext} ("Actual" value), V_{sup} ("Actual" value), I_{total} (FEG emission), I_{CT} (bottom right in the TstEGun window), I_{cup} (current meter), and the time in a table.

    1. Change from spot mode to scan mode.
    2. Reduce V_{ext} with 100 V.
    3. Change V_{sup} proportional to the V_{ext} change. This is to maintain the same field form around the tip (e.g., going from a 5.0 kV and 500 V to a 4.9 kV extraction voltage requires a suppressor voltage of 4.9/5.0 × 500 V).
    4. Set V_{C1} to the same value as the extractor.
    5. Set HV to the same value as the extractor.
    6. Set the condenser lens excitation to zero.
    7. If necessary focus a little bit on the aperture with the objective lens.
    8. Go to spot mode and put the beam inside the cup.
    9. Write down V_{ext}, V_{sup}, I_{total}, I_{CT}, I_{cup}, and the time in a table.

  - Repeat the above nine steps for the desired number of voltage settings.
  - Turn of the beam.
  - Bring back the extractor and the suppressor voltage to the settings of OPERATE mode (see the top part of your filled out source monitoring work sheet). **Wait until the values are set.**
  - Change ramping speeds of the extraction and the suppressor voltage from 10 to 0 (do not forget to press "set speeds").
  - Change from SERVICE mode to OPERATE mode.
  - Set back the aperture number from 1 to what it was before the measurement.
  - Beam on and check the system by making a high-resolution image.
  - Beam off.

- Vent the chamber.
- Take out the Faraday cup.
- Put the "insert" back into the objective lens, if required.
- Pump the chamber.
- Log off.
- Analyze the data and extract the field enhancement factor. Has it changed with respect to the last time, and what is the field strength the source is operating at?

A.2 Imaging of the Facet in the Quanta 200F System

The Quanta 200F extractor accepts part of the total facet beam: its acceptance half-angle is ~66 mrad (3.8°). If the C1 lens is turned off (same voltage as the extractor) or collimates the beam, the beam intensity profile in the CT behind the C1 lens is part of the facet emission pattern.

The intensity profile of the beam in the CT can be measured by scanning the beam across the CT aperture by using "gun tilt" and collecting the current through the aperture in a Faraday cup on the stage.

The gun tilt can be changed manually in the TstEOptics menu under "Gun tilt," with a maximum range of −1 to 1 for both x and y. The "Gun tilt" setting determines which part of the beam is selected by the CT aperture. This is done by changing the settings of the two sets of deflector plates, which are positioned somewhere along the CT. Because there are two sets of deflectors the pivot point can be maintained at the same height. This is done automatically when changing the gun tilt.

The divergence or collimation of the facet beam leaving the extractor can be tuned by changing the spot number (which changes the C1 lens setting): the part of the beam that is scanned by changing the gun tilt settings increases with increasing spot number. If the beam diameter at the CT aperture becomes smaller than the aperture size the recorded image will start showing the shadow image of the extractor aperture.

Appendix B

Procedure to Characterize System Performance

The performance of a probe-forming system can be characterized if the following parameters are available:

Source: practical brightness, FW50 energy spread, extraction voltage, virtual source size (B, ΔE, V_{ext}, d_v)
Column: spherical and chromatic aberration coefficients of the final lens (C_S, C_C)
Gun: spherical and chromatic aberration coefficients of the gun lens (C_{Sg}, C_{Cg})
Target: beam energy (V_p)

Key is which probe size contributions are dominant at the desired probe current level?

Approach: Calculate the parameters below and follow the scheme to find (1) which probe size contributions are dominant for your system at the probe current of interest, (2) the optimum probe size, (3) the associated half opening angle of the beam, and (4) the magnification of the virtual source.

Calculate $I_{AI} = 1.08 \cdot 10^{-18} B$ (I for which $d_A = d_I$)

Calculate $\delta_{CA} = 1.99 \cdot 10^{-5} \sqrt{C_C \dfrac{\Delta E}{V_p^{3/2}}}$ and $\delta_{SA} = 8.46 \cdot 10^{-8} \left(\dfrac{C_S}{V_p^{3/2}} \right)^{1/4}$

Calculate $I_{AC_g} = 2.72 \cdot 10^{-9} \dfrac{B V_{ext}^{3/2} d_v^2}{C_{Cg} \Delta E}$ and $I_{AS_g} = 1.50 \cdot 10^{-4} \dfrac{B V_{ext}^{3/4} d_v^2}{C_{Sg}^{1/2}}$

(I for which $d_A = d_{Cg}$) (I for which $d_A = d_{Sg}$)

Calculate $I_{\text{ICg}} = \dfrac{I_{\text{ACg}}^2}{I_{\text{AI}}}$ and $I_{\text{ISg}} = \left(\dfrac{I_{\text{ASg}}^2}{\sqrt{I_{\text{AI}}}}\right)^{2/3}$

(I for which $d_{\text{I}} = d_{\text{Sg}}$)

Is the desired probe current below I_{AI}?

$\delta_{\text{CA}} > \delta_{\text{SA}}$: d_{C} balances d_{A} → d_{p} from Table B.1; α and M from Table B.2
$\delta_{\text{CA}} < \delta_{\text{SA}}$: d_{S} balances d_{A}

Is the desired probe current between I_{AI} and the smallest of (I_{ISg},I_{ICg})?

$\delta_{\text{CA}} > \delta_{\text{SA}}$: d_{C} balances d_{I} → d_{p} from Table B.1; α and M from Table B.2
$\delta_{\text{CA}} < \delta_{\text{SA}}$: d_{S} balances d_{I}

Is the desired probe current larger than the smallest of (I_{ISg},I_{ICg})?

$I_{\text{ICg}} > I_{\text{ISg}}$:
$\delta_{\text{CA}} > \delta_{\text{SA}}$: d_{C} balances d_{Sg} → d_{p} from Table B.1; α and M from Table B.2
$\delta_{\text{CA}} < \delta_{\text{SA}}$: d_{S} balances d_{Sg}
$I_{\text{ICg}} < I_{\text{ISg}}$:

desired probe current smaller than I_{CgSg}?

$\delta_{\text{CA}} > \delta_{\text{SA}}$: d_{C} balances d_{Cg} → d_{p} from Table B.1; α and M from Table B.2
$\delta_{\text{CA}} < \delta_{\text{SA}}$: d_{S} balances d_{Cg}

desired probe current larger than I_{CgSg}?

d_{Cg} balances d_{Sg} → d_{p} from Table B.1; α and M from Table B.2

Table B.1 Optimum total probe size d_{p} for the desired probe current for a balance between d_{A}, d_{I}, d_{Cg} or d_{Sg}, and d_{C} or d_{S}. Also given is the figure of merit (fm) for column and source

	d_{C}			d_{S}		
	d_{p} (m)	fm column	fm source	d_{p} (m)	fm column	fm source
d_{A}	$\sqrt{2}\delta_{\text{CA}}$	$\dfrac{1}{C_{\text{C}}}$	$\dfrac{1}{\Delta E}$	$1.15\delta_{\text{SA}}$	$\dfrac{1}{C_{\text{S}}}$	-
d_{I}	$\sqrt{2}\delta_{\text{CA}}\left(\dfrac{I}{I_{\text{AI}}}\right)^{1/4}$	$\dfrac{1}{C_{\text{C}}}$	$\dfrac{B}{\Delta E^2}$	$1.54\delta_{\text{SA}}\left(\dfrac{I}{I_{\text{AI}}}\right)^{3/8}$	$\dfrac{1}{C_{\text{S}}}$	B

	d_C			d_S		
	d_p (m)	fm column	fm source	d_p (m)	fm column	fm source
d_{Cg}	$\delta_{CA}\left(\frac{I}{I_{ACg}}\right)^{1/2}$	$\frac{1}{C_C C_{Cg}}$	$\frac{V_{ext}^{3/2}Bd_v^2}{\Delta E^2}$	$\sqrt{2}\delta_{SA}\left(\frac{I}{I_{ACg}}\right)^{3/4}$	$\frac{1}{C_S C_{Cg}^{3/4}}$	$\frac{V_{ext}^{3/2}Bd_v^2}{\Delta E}$
d_{Sg}	$\sqrt{2}\delta_{CA}\left(\frac{I}{I_{ASg}}\right)^{1/1}$	$\frac{1}{C_C C_{Sg}}$	$\frac{V_{ext}^{3/4}Bd_v^2}{\sqrt{\Delta E}}$	$1.75\delta_{SA}\left(\frac{I}{I_{ASg}}\right)^{3/2}$	$\frac{1}{C_S C_{Sg}^{3/4}}$	$V_{ext}^{3/4}Bd_v^2$

Table B.2 Optimum magnification of the virtual source, M, and optimum half opening angle of the beam at the target, α_p, for different regimes in which two balancing contributions dominate the total probe size

	d_C		d_S	
	M	$\alpha_{p,opt}$ (rad)	M	$\alpha_{p,opt}$ (rad)
d_A	–	$\frac{6.62\cdot 10^{-10}}{\sqrt{V_p}\delta_{CA}}$	–	$\frac{6.18\cdot 10^{-10}}{\sqrt{V_p}\delta_{SA}}$
d_I	$\sqrt{2}\frac{d_p}{d_v}$	$\frac{6.62\cdot 10^{-10}}{\sqrt{V_p}\delta_{CA}}\left(\frac{I}{I_{AI}}\right)^{1/4}$	$0.80\frac{d_p}{d_v}$	$\frac{4.63\cdot 10^{-10}}{\sqrt{V_p}\delta_{SA}}\left(\frac{I}{I_{AI}}\right)^{1/8}$
d_{Cg}	$\left(\frac{C_C}{C_{Cg}}\right)^{1/2}\left(\frac{V_g}{V_p}\right)^{3/4}$	$\frac{6.62\cdot 10^{-10}}{\sqrt{V_p}\delta_{CA}}\left(\frac{I}{I_{ACg}}\right)^{1/2}$	$\frac{\delta_{SA}}{d_v}\frac{I_{ACg}^{1/4}}{\sqrt{I_{AI}}}I^{1/4}$	$\frac{6.62\cdot 10^{-10}}{\sqrt{V_p}\delta_{SA}}\left(\frac{I}{I_{ACg}}\right)^{1/4}$
d_{Sg}	$\frac{\delta_{CA}}{d_v}\frac{I_{ASg}}{\sqrt{I_{AI}}}\frac{1}{\sqrt{I}}$	$\frac{6.62\cdot 10^{-10}}{\sqrt{V_p}\delta_{CA}}\left(\frac{I}{I_{ASg}}\right)^{1/1}$	$\left(\frac{C_S}{C_{Sg}}\right)^{1/4}\left(\frac{V_g}{V_p}\right)^{3/8}$	$\frac{6.62\cdot 10^{-10}}{\sqrt{V_p}\delta_{SA}}\left(\frac{I}{I_{ASg}}\right)^{1/2}$

Bibliography

1. S. Babin, M. Gaevsky, D. Joy, M. Machin, A. Martynov, *J. Vac. Sci. Technol. B*, **24** (2006), 2956.
2. A. S. Bahm, G. A. Schwind, L. W. Swanson, *J. Vac. Sci. Technol. B*, **26** (2008), 2080.
3. J. P. Barbour, F. M. Charbonnier, W. W. Dolan, W. P. Dyke, E. E. Martin, J. K. Trolan, *Phys. Rev.*, **117** (1960), 1452.
4. J. E. Barth, P. Kruit, *Optik*, **101** (1996), 101.
5. J. E. Barth, M. D. Nykerk, *Nucl. Instrum. Methods, Phys. Res. A*, **427** (1999), 86.
6. A. E. Bell, L. W. Swanson, *Phys. Rev. B*, **19** (1979), 3353.
7. K. Besocke, H. Wagner, *Phys. Rev. B*, **8** (1973), 4597.
8. K. Besocke, B. Krahl-Urban, H. Wagner, *Surf. Sci.*, **68** (1977), 39.
9. P. C. Bettler, G. G. Love, *J. Appl. Phys.*, **38** (1967), 3835.
10. P. C. Bettler, G. Barnes, *Surf. Sci.*, **10** (1968), 165.
11. P. C. Bettler, D. H. Bennum, C. M. Case, *Surf. Sci.*, **44** (1974), 360.
12. H. Boersch, *Z. Phys.*, **139** (1954), 115.
13. C. Bombis, A. Emundts, M. Nowicki, H. P. Bonzel, *Surf. Sci.*, **511** (2002), 83.
14. H. P. Bonzel, *Phys. Rep.*, **385** (2003), 1.
15. A. S. Carasso, D. S. Bright, A. E. Vladár, *Opt. Eng.*, **41** (2002), 2499.
16. S. P. Chen, *Surf. Sci. Lett.*, **274** (1992), L619.
17. S. G. Christov, *Phys. Status Solidi*, **17** (1966), 11.
18. *Charged Particle Optics*, http://www.electronoptics.com.
19. A. C. Crewe, in *Handbook of Charged Particle Optics*, edited by J. Orloff (CRC, New York, 1997), Chap. 10, p. 406.

20. U. Dahmen, R. Erni, V. Radmilovic, C. Kisielowski, M.D. Rossell, P. Denes, *Philos. Trans. R. Soc. A*, **367** (2009), 3795.
21. L. R. Danielson, L. W. Swanson, *Surf. Sci.*, **88** (1979), 14.
22. L. R. Danielson, *J. Appl. Phys.*, **52** (1981), 6769.
23. M. Degawa, K. Thürmer, E. D. Williams, *Phys. Rev. B*, **74** (2006a), 155432.
24. M. Degawa, K. Thürmer, E. D. Williams, *Jpn. J. Appl. Phys.*, **45** (2006b), 2070.
25. A. K. Dokania, J. F. M. Velthuis, Y. Zhang, P. Kruit, *J. Vac. Sci. Technol. B*, **25** (2007), 504.
26. A.K. Dokania, M. Pelle, P. Kruit, *Microelectron. Eng.*, **85** (2008), 1031.
27. W. P. Dyke, F. M. Charbonnier, R. W. Strayer, R. L. Floyd, J. P. Barbour, J. K. Trolan, *J. Appl. Phys.*, **31** (1960), 790.
28. A. B. El-Kareh, J. C. Wolfe, J. E. Wolfe, *J. Appl. Phys.*, **48** (1977), 4749.
29. A. Ferriere, L. Lestrade, J-F. Robert, *J. Solar Energy Eng.*, **122** (2000), 9.
30. J. H. Fink, B. W. Schumacher, *Nucl. Instrum. Methods*, **130** (1975), 353.
31. R. G. Forbes, *J. Vac. Sci. Technol. B*, **26** (2008), 788.
32. M. J. Fransen, E. P. N. Damen, C. Schiller, T. L. van Rooy, H. B. Groen, P. Kruit, *Appl. Surf. Sci.*, **94–95** (1996), 107.
33. M. J. Fransen, J. S. Faber, T. L. van Rooy, P. C. Tiemeijer, P. Kruit, *J. Vac. Sci. Technol. B*, **16** (1998), 2063.
34. M. J. Fransen, T.L. de Rooy, P. C. Tiemeijer, M. H. F. Overwijk, J. S. Faber, P. Kruit, *Adv. Imag. Electron Phys.*, **111** (1999), 91.
35. M. J. Fransen, M. H. F. Overwijk, P. Kruit, *Appl. Surf. Sci.*, **146** (1999), 357.
36. S. Fujita, H. Shimoyama, *J. Electron Microsc.*, **54** (2005), 331.
37. S. Fujita, H. Shimoyama, *J. Electron Microsc.*, **54** (2005), 413.
38. S. Fujita, H. Shimoyama, *J. Vac. Sci. Technol. B*, **24** (2006), 1891.
39. S. Fujita, H. Shimoyama, *Phys. Rev. B*, **75** (2007), 235431.
40. G. Fursey, in *Field Emission in Vacuum Microelectronics*, edited by I. Brodie, P. Shwoebel, (Kluwer Academic/Plenum Publishers, New York, 2005), Chap. 7, p. 115.
41. M. Gesley, *J. Vac. Sci. Technol. B*, **9** (1991), 2949.
42. M. Giesen, G. Schulze Icking-Konert, H. Ibach, *Phys. Rev. Lett.*, **80** (1998), 552.
43. M. Giesen, *Prog. Surf. Sci.*, **68** (2001), 1.

44. M. Giesen, G. Beltramo, S. Dieluweit, J. Müller, H. Ibach, W. Schmickler, *Surf. Sci.*, **595** (2005), 127.

45. J. B. McGinn, S. G. den Hartog, D. S. Jun, G. G. Magera, G. A. Schwind, United States Patent 6680562 B1 (2004).

46. R. Gomer, L. W. Swanson, *J. Chem. Phys.*, **38** (1963), 1613.

47. D. J. Griffiths, *Introduction to Quantum Mechanics* (2nd ed.) (Prentice Hall, 2004), Chap. 8, p. 315.

48. G. A. Haas, in *The American Institute of Physics Handbook*, 3rd ed., edited by D. E. Gray (McGraw-Hill, 1972), pp. 9-172–9-186.

49. P. W. Hawkes, E. Kasper, *Principles of Electron Optics*, Vol. 2 (Academic Press, London, 1996), Chap. 47, p. 971.

50. P. W. Hawkes, E. Kasper, *Principles of Electron Optics*, Vol. 2 (Academic Press, London, 1996), Chap. 44, p. 924.

51. C. Herring, M. H. Nichols, *Rev. Mod. Phys.*, **21** (1949), 185.

52. C. Herring, in *The Physics of Powder Metallurgy*, edited by W. E. Kingston (McGraw–Hill, New York, 1951), Chap. 8, p. 143.

53. N. Hirai, K. Watanabe, S. Hara, *Surf. Sci.*, **493** (2001), 568.

54. M. Horn-Von Hoegen, F.-J. Meyer zu Heringdorf, D. Kähler, T. Schmidt, E. Bauer, *Thin Solid Films*, **336** (1998), 16.

55. W. Hu, X. Shu, B. Zhang, *Comput. Mater. Sci.*, **23** (2002), 175.

56. A. R. Hutson, *Phys. Rev.*, **98** (1955), 889; G. F. Smith, *Phys. Rev.*, **100** (1955), 1115.

57. H. Ibach, W. Schmickler, *Phys. Rev. Lett.*, **91** (2003), 016106.

58. H. Ibach, *Physics of Surfaces and Interfaces*, (Springer-Verlag, Berlin, Heidelberg, 2006), Chap. 4, p. 149.

59. A. Ichimiya, K. Hayashi, E. D. Williams, T. L. Einstein, M. Uwaha, K. Watanabe, *Appl. Surf. Sci.*, **175–176** (2001), 33.

60. Y. Irokawa, R. Mitsuhashi, S. Lee, K. Min, M. Inoue, Y. Kimura, R, Shimizu, *Jpn. J. Appl. Phys.*, **35** (1996), 4042.

61. N. Israeli, D. Kandel, *Phys. Rev. B*, **60** (1999), 5946.

62. S. C. Jain, V. Sinhas, B. K. Reddys, *J. Phys. D: Appl. Phys.*, **3** (1970), 1359.

63. G. H. Jansen, *Adv. Electron. Electron Phys.*, (Suppl. 21) (1990).

64. X. R. Jiang, P. Kruit, *Microelectron. Eng.*, **30** (1996), 249.

65. X. R. Jiang, *Coulomb Interactions in Charged Particle Optical Columns* (PhD thesis, Delft University of Technology, 1996), ISBN 90-407-1369-3.

66. N. de Jonge, *J. Appl. Phys.*, **95** (2004), 673.
67. D. C. Joy, *J. Microsc.*, **208** (2002), 24.
68. D. C. Joy, in *Characterization and Metrology for ULSI Technology 2005: Richardson, Texas, 15-18 March 2005*, edited by D. G. Seiler, Paper no. CP 788, (American Institute of Physics, Melville, 2005), p. 535.
69. N. K. Kang, D. Tuggle, L. W. Swanson, *Optik*, **63** (1983), 313.
70. H. Kawano, *Prog. Surf. Sci.*, **83** (2008), 1.
71. E. C. Kemble, *Phys. Rev.*, **48** (1935), 549.
72. H. S. Kim, M. L. Yu, E. Kratschmer, B. W. Hussey, M. G. R. Thomson, T. H. P. Chang, *J. Vac. Sci. Technol. B*, **13** (1995), 2468.
73. H. S. Kim, M. L. Yu, M. G. R. Thomson, E. Kratschmer, T. H. P. Chang, *J. Vac. Sci. Technol. B*, **15** (1997), 2284.
74. H. S. Kim, M. L. Yu, M. G. R. Thomson, E. Kratschmer, T. H. P. Chang, *J. Appl. Phys.*, **81** (1997), 461.
75. W. Knauer, *Optik*, **59** (1981), 335.
76. E. Kratschmer, S. A. Rishton, D. P. Kern, T. H. P Chang, *J. Vac. Sci. Technol. B*, **6** (1988), 2074.
77. P. Kruit, M. Bezuijen, J. E. Barth, *J. Appl. Phys.*, **99** (2006), 024315.
78. P. Kruit, G. H. Jansen, in *Handbook of charged particle optics (Second Edition)*, edited by J. Orloff (CRC Press, New York, 2008), Chap. 7, p. 341.
79. M. V. Kuz'min, M. A. Mittsev, *Tech. Phys. Lett.*, **27** (2001), 437.
80. N. D. Lang, W. Kohn, *Phys. Rev. B*, **1** (1970), 4555.
81. N. D. Lang, W. Kohn, *Phys. Rev.*, **3** (1971), 1215.
82. R. R. Law, *Proc. Inst. Radio Eng.*, **25** (1937), 954.
83. S. C. Lee, Y. Irokawa, M. Inoue, R. Shimizu, *Surf. Sci.*, **330** (1995), 289.
84. J. A. Liddle, P. Naulleau, G. Schmid, *J. Vac. Sci. Technol. B*, **22** (2004), 2897.
85. D.-J. Liu, J. D. Weeks, M. D. Johnson, E. D. Williams, *Phys. Rev. B*, **55** (1997), 7653.
86. D. Liu, H. M. Lu, Q. Jiang, *J. Phys.: Condens. Matter*, **21** (2009), 198002.
87. A. Y. Lozovoi, A. Alavi, *Phys. Rev. B*, **68** (2003), 245416.
88. J. Mackenzie, A. J. W. Moore, J. F. Nicholas, *J. Phys. Chem. Solids*, **23** (1962), 185.
89. T. Matsumoto, A. Cezairliyan, D. Basak, *Int. J. Thermophys.*, **20** (1999), 943.

90. T. Matsumoto, A. Ono, G. Barreiro, paper presented at the Fourteenth Symposium on Thermophysical Properties, June 25–30, 2000, Boulder, Colorado, USA.

91. I. Merrick, J. E. Inglesfield, G. A. Attard, *Phys. Rev. B*, **72** (2005), 033403.

92. A. Modinos, *Field, Thermionic, and Secondary Electron Emission Spectroscopy*, (Plenum Press, New York, 1984), Chap. 1, p. 1.

93. A. Modinos, J. P. Xanthakis, *Surf. Sci.*, **249** (1991), 373.

94. J. Müller, H. Ibach, *Phys. Rev. B*, **74** (2006), 085408.

95. I. Müllerová, *Scan. Microsc.*, **13** (1999), 7.

96. W. W. Mullins, G. S. Rohrer, *J. Am. Ceram. Soc.*, **83** (2000), 214.

97. E. L. Murphy, R. H. Good, *Phys. Rev.*, **102** (1956), 1464.

98. Y. Neo, H. Mimura, T. Matsumoto, *Appl. Phys. Lett.*, **88** (2006), 073511.

99. F. A. Nichols, W. W. Mullins, *J. Appl. Phys.*, **36** (1965), 1826.

100. J. Orloff, L. W. Swanson, J.-Z. Li, *J. Vac. Sci. Technol. B*, **3** (1985), 224.

101. V. G. Pavlov, *Phys. Solid State*, **48** (2006), 720.

102. I. Petroff, C. R. Viswanathan, *Phys. Rev. B*, **4** (1971), 799.

103. T. Radlicka, B. Lencová, *Ultramicroscopy*, **108** (2008), 455.

104. F. H. Read, N. J. Bowring, *Rev. Sci. Instrum.*, **74** (2003), 2280.

105. F. H. Read, N. J. Bowring, *Nucl. Instrum. Methods A*, **519** (2004), 196.

106. L. Reimer, *Scanning Electron Microscopy: Physics of Image Formation and Microanalysis* (Springer, Berlin, 1998), Chap. 2, p. 53.

107. S. A. Rishton, S. P. Beaumont, C. D. W. Wilkinson, *J. Phys. E: Sci. Instrum.*, 17 (1984), 296.

108. A. M. Rodriguez, G. Bozzolo, J. Ferrante, *Surf. Sci.*, **289** (1993), 100.

109. S. Sakawa, K. Tsunoda, Y. Terui, *Surf. Interface Anal.*, **35** (2003), 11.

110. N. Samoto, R. Shimizu, H. Hashimoto, N. Tamura, K. Gamo, S. Namba, *Jpn. J. Appl. Phys., 1*, **24** (1985), 766.

111. G. A. Schwind, D. S. Jun, G. G. Magera, United States Patent 6798126 B2, 2004.

112. G. A. Schwind, G. Magera, L. S. Swanson, *J. Vac. Sci. Technol. B*, **24** (2006), 2897.

113. H. Shimoyama, S. Maruse, *Ultramicroscopy*, **15** (1984), 239.

114. V. N. Shrednik, V. G. Pavlov, A. A. Rabinovich, B. M. Shaikhin, *Phys. Status Solids A*, **23** (1974), 373.

115. J. A. Simpson, *Rev. Sci. Instrum.*, **32** (1961), 1283.

116. H. L. Skriver, N. M. Rosengaard, *Phys. Rev. B*, **46** (1992), 7157.

117. D. F. Stafford, A. H. Weber, *J. Appl. Phys.*, **34** (1963), 2667.

118. *Surface Explorer*, http://surfexp.fhi-berlin.mpg.de/.

119. L. W. Swanson, G. A. Schwind, in *Handbook of Charged Particle Optics*, edited by J. Orloff (CRC Press, New York, 1997), Chap. 2, p. 77.

120. L. W. Swanson, G. A. Schwind, in *Handbook of Charged Particle Optics (Second Edition)*, edited by J. Orloff (CRC Press, New York, 2008), Chap. 1, p. 1.

121. A. Szczepkowicz, R. Bryl, *Surf. Sci.*, **559** (2004), L169.

122. M. Takeguchi, M. Morikawa, C. Hanqing, R. Shimizu, K. Tsunoda, H. Hagiwara, H. Hiraoka, *Technol. Rep. Osaka Univ.*, **41** (1991), 267.

123. K. Tamura, M. Amano, W. Chu, H. Ishii, M. Owari, T. Kawano, T. Nagatomi, Y. Takai, C. Oshima, R. Shimizu, Y. Nihei, *Surf. Interface Anal.*, **37** (2005), 217.

124. H. Tanaka, H. Nakayama, K. Watanabe, *Jpn. J. Appl. Phys.*, **44** (2005), 7518.

125. S. Tanaka, N. C. Bartelt, C. C. Umbach, R. M. Tromp, J. M. Blakely, *Phys. Rev. Lett.*, **78** (1997), 3342.

126. M. G. R. Thomson, *J. Vac. Sci. Technol. B*, **12** (1994), 3498.

127. K. Thürmer, J. E. Reutt-Robey, E. D. Williams, *Surf. Sci.*, **537** (2003), 123.

128. J. K. Trolan, J. P. Barbour, E. E. Martin, W. P. Dyke, *Phys. Rev.*, **100** (1955), 1646.

129. D. W. Tuggle (PhD thesis, Oregon Graduate Center, 1984).

130. M. Uwaha, *J. Phys. Soc. Jpn.*, **57** (1988), 1681.

131. A. H. V. van Veen (M.Sc. thesis, Delft University of Technology, 2000).

132. A. H. V. van Veen, C. W. Hagen, J. E. Barth, P. Kruit, *J. Vac. Sci. Technol. B*, **19** (2001), 2038.

133. L. Vitos, A. V. Ruban, H. L. Skriver, J. Kollár, *Surf. Sci.*, **411** (1998), 186.

134. A. E. Vladár, Z. Radi, M. T. Postek, D. C. Joy, *Scanning*, **28** (2006), 133.

135. W. R. Wade, *NASA-Memo-1-20-59L* (1959).

136. K. Wang, R. R. Reeber, *Mater. Sci. Eng.*, **R23** (1998), 101.

137. J. W. Ward, R. L. Kubena, M. W. Utlaut, *J. Vac. Sci. Technol. B*, **6** (1988), 2090.

138. Y.-N. Wen, J.-M. Zhang, *Comput. Mater. Sci.*, **42** (2008), 281.

139. J. C. Wiesner, T. E. Everhart, *J. Appl. Phys.*, **44** (1973), 2140.

140. E. D. Williams, *Surf. Sci.*, **299/300** (1994), 502.

141. J. Worster, *Brit. J. Appl. Phys. (J. Phys. D)*, **2** (1969), 889.

142. W. Xu, J. B. Adams, T. L. Einstein, *Phys. Rev. B*, **54** (1994), 2910.

143. R. Young, S. Henstra, J. Chmelik, T. Dingle, A. Mangnus, G. van Veen, I. Gestmann, *Proc. SPIE*, **7378** (2009), 737803.

144. J.-M. Zhang, F. Ma, K.-W. Xu, *Surf. Interface. Anal.*, **35** (2003), 662.

145. Y. Zhang, P. Kruit, *J. Vac. Sci. Technol. B*, **25** (2007), 2239.

146. T. M. Gardiner, H. M. Kramer, E. Bauer, *Surf. Sci.*, **112** (1981) 181.

List of publications associated with this work:

- M. S. Bronsgeest, P. Kruit, *J. Vac. Sci. Technol. B*, **24** (2006), 887: "Temperature dependence of the work function of the Zr/O/W (100), Schottky electron source in typical operating conditions and its effect on beam brightness" (Section 3.5.1 of this book).
- M. S. Bronsgeest, J. E. Barth, G. A. Schwind, L. W. Swanson, P. Kruit, *J. Vac. Sci. Technol. B*, **25** (2007), 2049: "Extracting the Boersch effect contribution from experimental energy spread measurements for Schottky electron emitters" (Section 4.4.2 of this book).
- M. S. Bronsgeest, J. E. Barth, L. W. Swanson, P. Kruit, *J. Vac. Sci. Technol. B*, **26** (2008), 949: "Probe current, probe size, and the practical brightness for probe forming systems" (Section 4.2 of this book).
- M. S. Bronsgeest, P. Kruit, *J. Vac. Sci. Technol. B*, **26** (2008), 2073: "Effect of the electric field on the form stability of a Schottky electron emitter: a step model" (Section 5.5.1 of this book).
- M. S. Bronsgeest, P. Kruit, *J. Vac. Sci. Technol. B*, **27** (2009), 2524: "Reversible shape changes of the end facet on Schottky electron emitters" (Sections 5.4.2 and 5.6.1 of this book).
- M. S. Bronsgeest, P. Kruit, *Ultramicroscopy*, **110** (2010), 1243: "'Collapsing rings' on Schottky electron emitters" (Sections 5.5.2–5.6.2 of this book).

Index